Genome Editing in Bacteria

(Part 1)

Edited By

Prakash M. Halami

Department of Microbiology and Fermentation Technology
CSIR- Central Food Technological Research Institute
Mysuru-570020
India

Academy of Scientific and Innovative
Research (AcSIR), Ghaziabad
Uttar Pradesh, India

&

Aravind Sundararaman

Department of Microbiology and Fermentation Technology
CSIR- Central Food Technological Research Institute
Mysuru-570020
India

Genome Editing in Bacteria (Part 1)

Editors: Prakash M. Halami and Aravind Sundararaman

ISBN (Online): 978-981-5165-67-8

ISBN (Print): 978-981-5165-68-5

ISBN (Paperback): 978-981-5165-69-2

need for a court order if at any point you breach any terms of this License Agreement. In no event will any delay or failure by Bentham Science Publishers in enforcing your compliance with this License Agreement constitute a waiver of any of its rights.

3. You acknowledge that you have read this License Agreement, and agree to be bound by its terms and conditions. To the extent that any other terms and conditions presented on any website of Bentham Science Publishers conflict with, or are inconsistent with, the terms and conditions set out in this License Agreement, you acknowledge that the terms and conditions set out in this License Agreement shall prevail.

Bentham Science Publishers Pte. Ltd.
80 Robinson Road #02-00
Singapore 068898
Singapore
Email: subscriptions@benthamscience.net

BENTHAM SCIENCE

CONTENTS

FOREWORD

The importance of Biotechnology and its applications in our life is well known, and with nearly four decades of experience at the University level, I am delighted and glad to write the foreword for the book "Genome Editing in Bacteria" edited by Dr. Prakash Halami and Dr. Aravind Sundararaman. I had known Dr. Halami since 2008, when I was nominated to the Institutional Biosafety Committee of CSIR-CFTRI, Mysuru, by the Dept. of Biotechnology, Govt. of India (New Delhi). I appreciate Dr. Halami abilities in planning, executing, monitoring research projects and dedication in mentoring many researchers in his lab.

Biotechnology is one of the most significant branches of biological sciences that is shaping the century and will continue to flourish and expand to newer frontiers in the present century too. Biotechnology draws parallel developments through research and applications in Microbiology, Genetics, Molecular biology, Bioinformatics, and Nanotechnology. Diverse branches of biotechnology have distinguished niches to well-defined dynamic research areas with both academic and industrial applications. Genome editing is a subject that has turned into a high science topic in our everyday vocabulary over a short period of time. Several positive and negative attributes have been associated with gene delivery techniques to develop transgenic microbes, animals and crops. The field has wide applications, and with every new development reported in leading peer-reviewed journals across the globe, the opportunities only become wider and the hopes brighter.

The editors have done an extraordinary job of bringing out a timely peer-reviewed volume titled "Genome Editing in Bacteria" with contributors spread across different continents. It is quite impressive to note that the editors have identified a wide range and dynamic topics that are organized in the form of a series of captivating articles highlighting different aspects of genetic engineering, both traditional and modern technologies in the field, protocols, advantages, new school of thoughts from around the world associated with some frontier development of biotechnological research. The attractive illustrations for presenting complex theoretical and experimental details and overall production design are sure to win the hearts of enthusiastic readers. The simplicity of the language and presentation style appealed to me a lot.

With years of experience in specific fields, several scientists and researchers have compiled and provided authentic and contemporary information. This compendium will be handy for faculty to update their lectures for the students pursuing higher education. Similarly, this book will be a ready reckoner for researchers working in specific areas to plan their future research.

It is a great pleasure on my part to pen a foreword for this prestigious, multi-authored, peer-reviewed, international publication on a topic that is very contemporary and close to my heart.

I wish the editors great success. I also extend my sincere greetings to all the contributors for their excellent effort in making this book a success.

S. R. Ramesh
Department of Studies in Zoology
University of Mysore
Mysore, India

PREFACE

Genetic engineering is essential in basic research and Industrial Biotechnology for metabolic and genomic manipulations to regulate microorganisms to produce valuable products. Genome editing is the cornerstone for scientists to interrogate the genetic basis of microorganisms for which the accessibility of the genome along with molecular tools is an essential factor. The classical genetic methods developed for genome editing in bacterial species include culture and transformation. The methods were highly laborious and required the introduction of at least one resistance marker cassette in the genome, which hampers the possibility of producing precise edits like single amino acid mutations. The breakthrough by the discovery of the CRISPR-Cas technology has shed light on the adaptive immune system of prokaryotes to explore tremendous opportunities for targeted genetic engineering approaches in prokaryotes. Here, we discuss the current state-of-the-art approach for gene editing in bacteria and different strategies used in this technology for prokaryotic organisms.

This book has eight chapters, including historical perspectives of genome editing and applications of probiotics and its metabolites. We attempted to update and collate information and research carried out on various applications of bacteria in different industries, such as the food and pharmaceutical industry and their gene regulation for metabolic engineering using genome editing tools. We are grateful to all contributing authors who accepted our invitation to contribute to this book. The contributing authors are well-recognized scientists and researchers with vast experience in the field of bacteriology and molecular biology. We are happy to bring them all together on the same platform to bring out this book. We are grateful to the Bentham Science Group for publishing this comprehensive book, and we hope it will be read by researchers, students, teachers, scientists and food entreprenuers who are interested in the metabolic engineering of bacteria for various health benefits. Although there are hundreds of research articles, review papers, and limited books on genome editing of prokaryotes, this book "Genome Editing in Bacteria" is the first of this kind, a compilation of various applications of bacteria across diverse fields of biotechnology.

We dedicate this book to the creators of the indigenous knowledge of molecular biology and genetic engineering for putting together both an ocean of knowledge and the basis for research to study in-depth genome editing techniques for bacteria.

Prakash M. Halami
Department of Microbiology and Fermentation Technology
CSIR- Central Food Technological Research Institute
Mysuru-570020
India

Academy of Scientific and Innovative
Research (AcSIR), Ghaziabad
Uttar Pradesh, India

&

Aravind Sundararaman
Department of Microbiology and Fermentation Technology
CSIR- Central Food Technological Research Institute
Mysuru-570020
India

List of Contributors

Angelina Job Kolady   Department of Genomic Science, School of Biological Sciences, Central University of Kerala, Tejaswini Hills, Periya (PO) Kasaragod, Kerala 671320, India

Aritra Mukherjee   Department of Genomic Science, School of Biological Sciences, Central University of Kerala, Tejaswini Hills, Periya (PO) Kasaragod, Kerala 671320, India

Ashif Ali   Biotechnology Division, CSIR-Institute of Himalayan Bioresource Technology Palampur, Himachal Pradesh-176061, India

Aman Kumar   Biotechnology Division, CSIR-Institute of Himalayan Bioresource Technology Palampur, Himachal Pradesh-176061, India
Academy of Scientific and Innovative Research (AcSIR), Ghaziabad-201002, India

Atul R. Chavan   Academy of Scientific and Innovative Research (AcSIR), Ghaziabad-201002, India
Environmental Biotechnology and Genomics Division (EBGD), CSIR–National Environmental Engineering Research Institute (NEERI), Nehru Marg, Nagpur-440020, India

Anshuman A. Khardenavis   Academy of Scientific and Innovative Research (AcSIR), Ghaziabad-201002, India
Environmental Biotechnology and Genomics Division (EBGD), CSIR–National Environmental Engineering Research Institute (NEERI), Nehru Marg, Nagpur-440020, India

Aravind Sundararaman   Department of Microbiology and Fermentation Technology, CSIR- Central Food Technological Research Institute, Mysuru-570020, India

Amit Kumar Rai   National Agri-Food Biotechnology Institute (DBT-NABI), S.A.S. Nagar, Mohali, Punjab, India

A. Sajna   Department of Biotechnology, Nehru Arts and Science College, Nehru Gardens, T M Palayam, Coimbatore, Tamil Nadu, India

Hemant J. Purohit   Environmental Biotechnology and Genomics Division (EBGD), CSIR–National Environmental Engineering Research Institute (NEERI), Nehru Marg, Nagpur-440020, Maharashtra, India

Kriti Ghatani   Department of Food Technology, University of North Bengal, Raja Rammohunpur, Darjeeling, West Bengal, 734013, India

Kiran Dindhoria   Biotechnology Division, CSIR-Institute of Himalayan Bioresource Technology Palampur, Himachal Pradesh-176061, India
Academy of Scientific and Innovative Research (AcSIR), Ghaziabad-201002, India

Loreni Chiring Phukon   National Agri-Food Biotechnology Institute (DBT-NABI), S.A.S. Nagar, Mohali, Punjab, India

Maitreyee Pathak   Environmental Biotechnology and Genomics Division (EBGD), CSIR–National Environmental Engineering Research Institute (NEERI), Nehru Marg, Nagpur-440020, India

Md Minhajul Abedin National Agri-Food Biotechnology Institute (DBT-NABI), S.A.S. Nagar, Mohali, Punjab, India

Pasupuleti Sreenivasa Rao Central Research Laboratory (ARC), Narayana Medical College and Hospital, Nellore-524003, India
Narayana College of Pharmacy, Nellore-524003, Andhra Pradesh, India

Priya Chakraborty Department of Food Technology, University of North Bengal, Raja Rammohunpur, Darjeeling, West Bengal, 734013, India

Puja Sarkar Institute of Bioresources and Sustainable Development, Regional Centre, Tadong, Sikkim, India

Prakash M. Halami Department of Microbiology and Fermentation Technology, CSIR- Central Food Technological Research Institute, Mysuru–570020, India
Academy of Scientific and Innovative Research (AcSIR), Ghaziabad, Uttar Pradesh, India

Rounak Chourasia Institute of Bioresources and Sustainable Development, Regional Centre, Tadong, Sikkim, India

Ranjith Kumavath Department of Genomic Science, School of Biological Sciences, Central University of Kerala, Tejaswini Hills, Periya (PO) Kasaragod, Kerala 671320, India
Department of Biotechnology, School of Life Science, Pondicherry University, Puducherry-605014, India

Rakshak Kumar Biotechnology Division, CSIR-Institute of Himalayan Bioresource Technology Palampur, Himachal Pradesh-176061, India
Academy of Scientific and Innovative Research (AcSIR), Ghaziabad-201002, India

Sarvepalli Vijay Kumar Department of Paediatrics, Lincoln University College, Kuala Lumpur, Malaysia

Shankar Prasad Sha Department of Botany, Food Microbiology Lab, Kurseong College, University of North Bengal, Dow Hill Road, Kurseong, Darjeeling 7342003, West Bengal, India

Subarna Thapa Department of Food Technology, University of North Bengal, Raja Rammohunpur, Darjeeling, West Bengal, 734013, India

Sagnik Sarkar Department of Food Technology, University of North Bengal, Raja Rammohunpur, Darjeeling, West Bengal, 734013, India

Srichandan Padhi Institute of Bioresources and Sustainable Development, Regional Centre, Tadong, Sikkim, India

Sudhir P. Singh Center of Innovative and Applied Bioprocessing (DBT-CIAB), S.A.S. Nagar, Mohali, India

Sudeepa E. S. Department of Biotechnology, Nehru Arts and Science College, Nehru Gardens, T M Palayam, Coimbatore, Tamil Nadu, India

Steji Raphel Department of Microbiology & Fermentation Technology, CSIR- Central Food Technological Research Institute, Mysuru–570020, India
Academy of Scientific and Innovative Research (AcSIR), Ghaziabad, Uttar Pradesh, India

Vivek Manyapu   Biotechnology Division, CSIR-Institute of Himalayan Bioresource Technology
Palampur, Himachal Pradesh-176061, India

CHAPTER 1

Recent Advances in CRISPR-Cas Genome Engineering: An Overview

Angelina Job Kolady[1], Aritra Mukherjee[1], Ranjith Kumavath[1,2,*], Sarvepalli Vijay Kumar[3] and Pasupuleti Sreenivasa Rao[4,5,*]

[1] *Department of Genomic Science, School of Biological Sciences, Central University of Kerala, Tejaswini Hills, Periya (PO) Kasaragod, Kerala 671320, India*

[2] *Department of Biotechnology, School of Life Science, Pondicherry University, Puducherry-605014, India*

[3] *Department of Paediatrics, Lincoln University College, Kuala Lumpur, Malaysia*

[4] *Central Research Laboratory (ARC), Narayana Medical College and Hospital, Nellore-524003, India*

[5] *Narayana College of Pharmacy, Nellore-524003, Andhra Pradesh, India*

Abstract: Bacteria is one of the most primitive organisms on earth. Its high susceptibility to bacteriophages has tailored them to use specific tools to edit their genome and evade the bacteriophages. This defense system has been developed to be the most specific genome editing technology of this current period. Previously, various other tools such as restriction enzymes (RE), zinc finger nucleases (ZNF), and transcription activator-like effector nucleases (TALENS) were utilized. Still, its major limitations led to exploiting the bacterial defense system to edit the genome. CRISPR technology can be applied in various microbiology, pathology, cancer biology, molecular biology, and industrial biotechnology, but its limitations, such as off-target effects due to unspecific alterations, are a major concern. In the future, this effective gene alteration technology will be developed to treat inherited rare genetic disorders. This chapter highlights the discovery, components, applications, limitations, and future prospects of CRISPR-Cas.

Keywords: Bacterial defense system, Cas9, CRISPR-Cas, Genome editing tools, Industrial biotechnology, SgRNA.

---

* **Corresponding authors Ranjith Kumavath and Pasupuleti Sreenivasa Rao:** Department of Genomic Science, School of Biological Sciences, Central University of Kerala, Tejaswini Hills, Periya (PO) Kasaragod, Kerala 671320, India; & Central Research Laboratory (ARC), Narayana Medical College and Hospital, Nellore-524003, India; Tel: +918547648620, E-mails: RNKumavath@gmail.com, RNKumavath@cukerala.ac.in

INTRODUCTION

Bacteria are single-celled prokaryotic microorganisms, all belonging to the kingdom of Monera in the system of classification of living organisms [1]. Bacteria are among the oldest living organisms as they were among the first life forms to appear on earth and are present in almost every habitat. We usually associate bacteria with an infectious disease. Nonetheless, every bacterium lives in parasitic relations with plants and animals. A significant number of bacterial species live in symbiotic associations with other living organisms. Bacteria are also prone to infection from specialized viruses called bacteriophages [2]. To evade conditions from bacteriophages, bacteria evolved to use a specialized tool in their genome and clustered regularly interspaced short palindromic repeats [3, 4].

GENOME EDITING

Genome editing is a process where specific changes can be made in the regions of interest with the help of explicit and engineered nucleases by introducing double-stranded breaks (DSB). These breaks can cause site-specific mutations, gene deletions, substitutions, or insertions, and later can be repaired by various mechanisms. Non-homologous end joining (NHEJ) is prone to error, and homology-directed repair (HDR) error-free is the repair mechanism used [5]. Genome editing is a powerful tool for understanding biological roles. It can treat genetic disorders by identifying 'molecular mistakes' and providing appropriate gene therapy. Restriction enzymes (RE) are natural genome editing tool, while transcription activator-like effector nucleases (TALENs) and Zinc Finger Nucleases (ZFNs) are artificial genome editing tools.

GENOME EDITING TOOLS AND THEIR LIMITATIONS

Restriction Enzymes (RE)

The discovery of restriction enzymes in early 1970 heralded a new age in molecular biology. Restriction enzymes or endonucleases are natural genome editing tools that recognize specific nucleotide sequences and cut the DNA at specific sites. The gene of interest could be inserted at a particular location.

The limitation of the restriction enzymes is the difficulty in predicting the location at which the gene of interest could be inserted. The primary reason behind it is that the recognition sequence of most of the restriction enzymes is base pairs long and often arises several times in a genome. The restriction specificity of endonucleases can depend on the environmental conditions.

In contrast, restriction enzymes are used for molecular cloning, DNA mapping, epigenome mapping, and constructing DNA libraries. These enzymes were modified to enhance the specificity of restriction endonucleases like the homing endonuclease systems. They could target specific sequences for genome editing. REs have long recognition sites and tolerate sequence degeneracy within their restriction site, unlike restriction enzymes [6]. One of the examples is meganucleases. It is designed to recognize long DNA sequences.

Zinc Finger Nucleases (ZNFs)

The artificial restriction enzymes consist of a subunit that recognizes desired DNA sequence and the DNA cutting part of restriction enzymes. They can be designed to identify specific DNA sequences and thereby enable targeted cleavage [7]. The hybrid restriction enzymes could be created using a zinc finger DNA binding domain fused to break up the naturally occurring FokI endonuclease domain. FokI, a naturally occurring IIS restriction enzyme, has played a pivotal role in the success of ZFNs. A lot of effort was required to produce a modified ZNF, which was a significant drawback. Thus, research has been done to customize ZNF.

Transcription Activator Like Effector Nucleases

The artificial restriction enzymes (ARE) consist of two components (i) restriction enzyme to cleave DNA and (ii) TAL, effector. TAL effectors comprise 33 repeat sequences, which helps them bind to long lines in the genome. TALENs are preferable over ZNF due to their ease of application. TALENs encode the FokI domain fused to the engineered DNA binding region, and when bound, dimerized FokI endonuclease could form a double-stranded break. The limitation of TALEN is TALE target search process is affected by genomic occlusions.

Discovery and History

The discovery of CRISPR revolutionized gene-editing technology (Fig. **1**). CRISPR was initially discovered in 1987 from the *E. coli* genome, and its role in the adaptive immune system was elucidated in early 2000. In 2020, Prof. Emmanuelle Charpentier and Prof. Jennifer Doudna were honored with the Nobel Prize in chemistry for their discovery of CRISPR-Cas9 technology in *Streptococcus pyogenes,* which is considered an evolution in the fields of medicine, biotechnology, and agriculture [8]. This technique is favorable due to its precision in gene editing [9]. This helps to alter genes efficiently and rapidly. It is also widely used in treating genetic disorders [10].

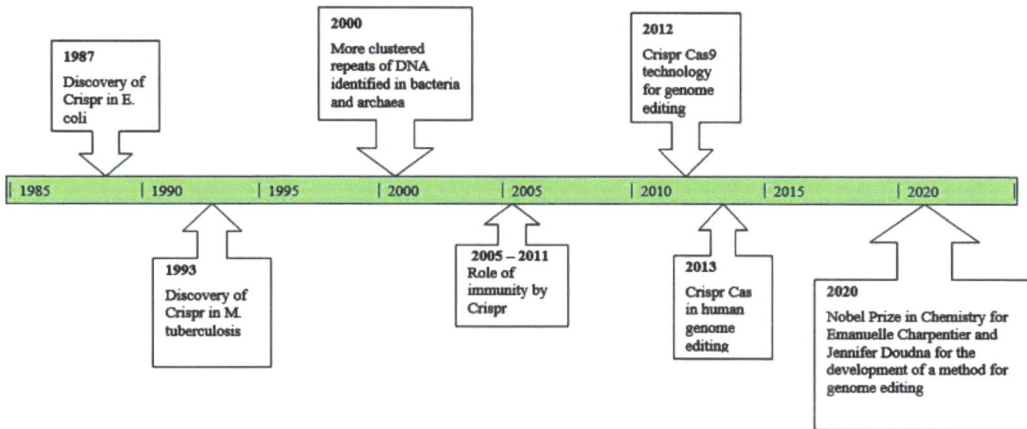

Fig. (1). Timeline of the discovery of CRISPR as the genome-editing tool. CRISPR was discovered in 1987 in *E. coli* and was later found in other species, from 2005 to 2011.

Spacer Acquisition

Upon invasion of a prokaryote phage virus, the first stage of the immune response is to capture phage DNA and insert it into the CRISPR locus in a spacer. New spacers are usually added upstream of the CRISPR next to the leader sequence creating chronological order of the viral infections. *Cas1* and *Cas2* are found in both CRISPR-Cas immune systems, hinting that they are involved in spacer acquisition [11 - 15]. However, their crystal structures are similar, and purified Cas1 proteins are metalloenzymes acting as nucleases/integrates that bind to DNA sequence-independent [16]. Representative Cas2 proteins have been characterized and contain either ssRNA (single strand) or dsDNA specific endoribonuclease activity. The analysis of CRISPR-Cas systems showed PAMs as necessary for type I and type II but not for type III systems during spacer acquisition [17]. The conservation of the PAM sequences differs between CRISPR-Cas systems and appears to be evolutionarily linked to Cas1 and the leader sequence [17].

Biogenesis

Specialized Cas proteins to form crRNAs then cleave this transcript. However, the type I-E and type I-F systems, proteins Cas6e and Cas6f, respectively, recognize stem-loops created by pairing identical repeats that flank the crRNA [18]. The cleavage instead occurs by the longer transcript enveloping around the Cas6 to allow detachment just upstream of the repeat sequence [19]. Type-II systems lack the Cas6 gene and instead use RNaseIII for cleavage. Functional type II systems encode a specialized form of RNA known as a trans-activating crRNA (tracrRNA) [20]. Transcription of the tracrRNA and the immediate CRISPR transcript results in base pairing and dsRNA formation at the repeat sequence, which RNaseIII eventually targets to produce crRNAs. The crRNA only contains a truncated spacer at one end, unlike the other two systems. The type I-E Cascade requires five Cas proteins bound to a single crRNA [21].

Components of CRISPR-Cas

Cas9

It is an endonuclease enzyme used to cut DNA at the target sequence. Cas9 can recognize and bind to the target sequence in front of the adjacent protospacer motif (PAM) sequence to enhance the specificity. Cas9 cuts at the specific site and causes a double-stranded break (DSB), after which the DNA repairing mechanism such as non-homology end joining (NHEJ) and homology-directed repair (HDR) occurs. One such example is SpyCas9 cDNA is found in several plasmids, especially pX330.

sgRNA and How to Design sgRNAs

The complement the DNA sequences on either side of the cut and contain whatever arrangement is desired for insertion into the host genome [22]. They occur naturally, serving essential functions, but can also be designed to be used for targeted editing, such as with CRISPR-Cas9 [23]. Most prokaryotes, encompassing bacteria and archaea, use CRISPR with its associated Cas enzymes as their adaptive immune system. When prokaryotes are infected by phages and manage to fend off the attack, Cas enzymes will cut phage DNA or RNA and integrate parts between the repeats of the CRISPR sequence [22].

Therefore, the trans-activating RNA (tracrRNA) and crRNA are two key components joined by tetraloop that results in the formation of sgRNA [24], which identifies the specific complementary target region, which is cleaved by Cas9 after binding with crRNA and tRNA, which are altogether known as effector complex. With the modifications in the rRNA sequences of the guide RNA, the binding location can be changed, hence defining it as a user-defined program [25], and finding 3-4 nucleotides downstream from the cut site. After base pairing of the gRNA to the target, Cas9 mediates a double-strand break about 3-nt upstream of PAM [23]. CRISPR/Cas9 is used for gene editing and gene therapy [25]. The evidence shows that both *in-vitro* and *in-vivo* required tracrRNA for Cas9 and target DNA sequence binding [26]. Several different proteins, like Cas1 and Cas2, help to find new spacers [22]. The Cas9 protein binds to a combined form of crRNA and tracrRNA, forming an effector complex—this guides RNA for the Cas9 protein directing it for its endonuclease activity [24].

PAM Sequence

A short DNA sequence (2-6 bp) follows the CRISPR system's DNA region targeted for cleavage. This sequence helps the Cas nuclease to cut at the specific target site. While genome editing, the presence of the PAM sequence, assists in locating the target sequence. The various Cas endonucleases recognize different PAM, which is advantageous during genome editing (Table 1). While designing gRNA, researchers exclude PAM from gRNA because bacteria exclude the PAM sequences to ensure that cuts do not occur on the bacterial genome.

Table 1. Summary of Cas nucleases, PAM sequences, and the organisms from which it was isolated.

CRISPR Nucleases	Organism Isolated from	PAM Sequence (5' to 3')
SpCas9	*Streptococcus pyogenes*	NGG
SaCas9	*Staphylococcus aureus*	NGRRT or NGRRN
NmeCas9	*Neisseria meningitidis*	NNNNGATT
CjCas9	*Campylobacter jejuni*	NNNNRYAC
StCas9	*Streptococcus thermophilus*	NNAGAAW

Component CRISPR

This is the most basic form of CRISPR used in mice and consists of two components, namely, Cas9 and sgRNA [27] (Fig. 2). For example, in 2C CRISPR, SpyCas9 and sgRNA resulting in an altered reading frame or a premature termination codon. DSB is repaired through error-prone non-homologous end joining (NHEJ), as it is the dominant mode of repair in mouse zygotes [5]. This can disrupt the reading frame. Deleting promoter elements leads to loss or silencing of gene expression. The applications of 2C CRISPR are the ease in knocking out non-coding genes. It is easy and widely used. Multiple gene loci can be targeted using 2C CRISPR, but the unpredictable nature of editing lies in the weakness [5].

Component CRISPR (3C CRISPR)

3C CRISPR combines a donor DNA template with Cas9 and sgRNA to promote homology-directed repair (HDR) over non-homologous end joining (NHEJ) after the double-stranded break (DSB) [5]. Nevertheless, the NHEJ is an error-prone mechanism in which broken ends of DNA are joined together, often resulting in a heterogeneous pool of insertions and deletions [28]. The most crucial application of 3C CRISPR is its precise and subtle genome editing (Fig. 3). It is used to correct cataracts by repairing a disrupted reading frame in the Crygc gene [29]. Since it can target exons efficiently, point mutations of various genetic disorders can be introduced, and expressions can be studied. The limitation of 3C CRISPR is the lower efficiency of repair in mice than NHEJ because HDR is confined to the replication phase of the cell cycle [5].

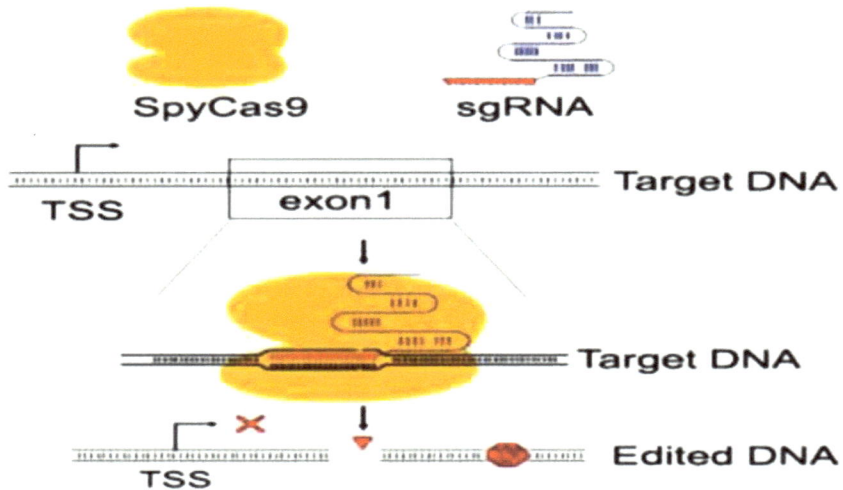

Fig. (2). 2C CRISPR-SpyCas9 and sgRNA are the two components involved in 2C CRISPR.

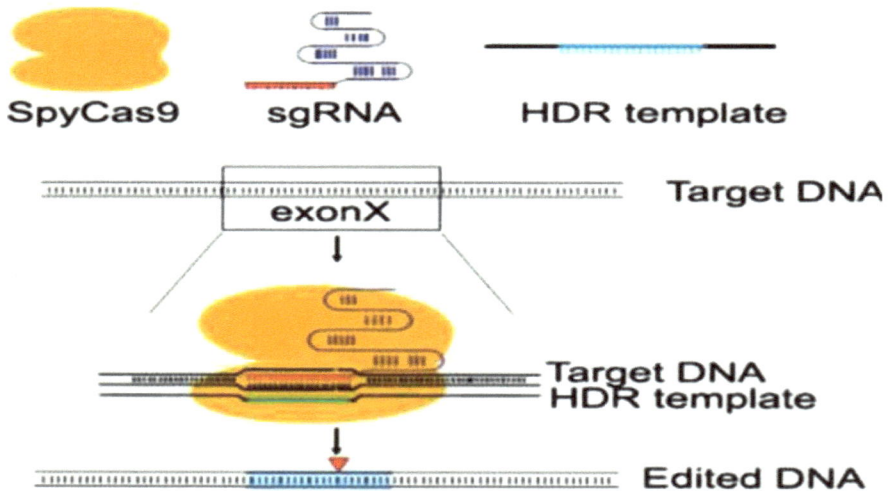

Fig. (3). 3C CRISPR– A donor template called the HDR template is chosen to enhance the specificity for the restriction endonuclease, such as SpyCas9 can cleave at specific sites. 3 C CRISPR is used to promote homology-directed repair rather than NHEJ (Adapted from "Joseph Miano *et al.* 2016").

APPLICATION OF CRISPR

• CRISPR can be used in agriculture, molecular biology, genetics, medicine, and others [30].

• It is used to treat mutations in genetic disorders. For example, rhodopsin mutation in autosomal dominant retinitis pigmentosa was correlated in rats by employing subretinal injection of plasmids encoding CRISPR-Cas9 and suitable sgRNAs [31].

• They also possess antiviral and antimicrobial properties.

• CRISPR is used to block phage infection in bacteria [32]. Human viruses such as HIV are targeted, and their entry can also be prevented.

• It plays a vital role in generating knockout and treating cancer patients [33].

Generation of Knockout

CRISPR is used to generate knockout cells or animals by co-expressing an endonuclease like Cas9 or Cas12a and a gRNA specific to the targeted gene (Fig. **4**). The target sequence should be adjacent to the PAM sequence as PAM helps Cas9 cut at the target site. They once expressed, Cas9 and gRNA from a ribonucleoprotein complex through interactions between the gRNA scaffold and surface-exposed positively charged grooves on Cas9. The binding of gRNA and Cas9 causes a conformational change. gRNA spacer sequence allows Cas9 for cleavage at the target site. Cas9-Gran complex binds to the putative DNA target, and annealing occurs. Cas9 undergoes a conformational change upon target binding, where nuclease RuvC and HNH cause double-stranded breaks, followed by the repair mechanism either by NHEJ or HDR. NHEJ is the most efficient and active mechanism that can cause indels, resulting in the loss of function mutation gene.

This knockout generation mechanism is also used to knock out cancer cell lines [34]. CRISPR-Cas9 could enable albumin production using transgenic pigs [35].

Making Specific Modifications

The genome modification in the Adult Rat Brain using CRISPR-Cas9 transgenic Rats, cell-specific and sequence-specific genome modifications in the adult brain could be made. The tyrosine hydroxylase (*Th*) gene, expressed in dopaminergic neurons of the midbrain, was targeted. gRNAs were constructed to bind to the first exon of the *Th* gene. Cas9 variants have been generated that have the ability to bind RNA and DNA. This has modified catalytic sites and can produce a nick in one strand of DNA [36, 37]. CRISPR-Cas9 is used to modify normal bone marrow hematopoietic stem and progenitor cells (HSPCs), which has led to a new approach to autologous transplantation therapy for the treatment of homozygous beta-thalassemia and sickle cell anemia (SCD) [38].

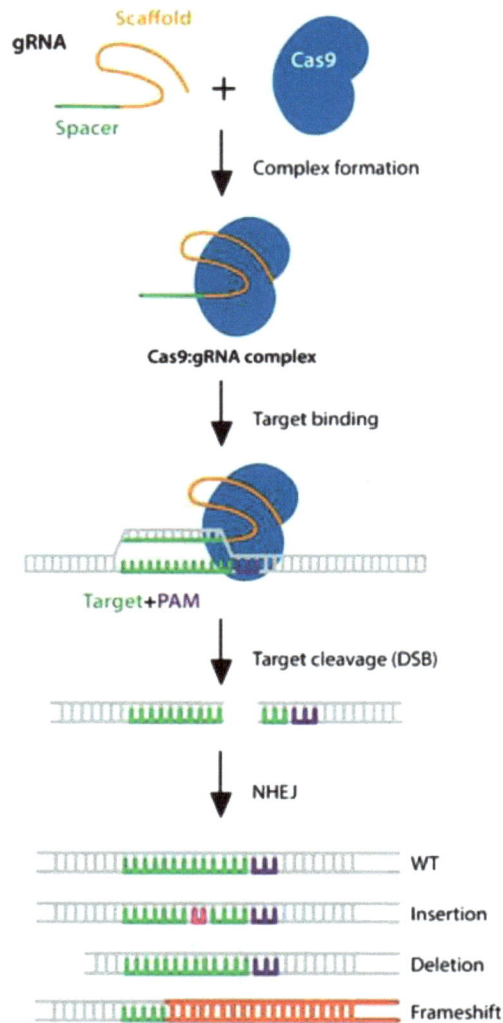

Fig. (4). Generation of knockout. In this process, Cas9 and gRNA form a complex, and with the pam sequence, a specific sequence of the DNA is targeted and cleaved. A double-stranded break is included, which is then repaired by non-homologous end-joining. Insertions, deletions, and substitutions can be introduced using 3C CRISPR (Adapted from www.addgene.org).

Combat Antimicrobial Resistance

One of the essential characteristics of CRISPR-Cas is its ability to invade foreign genetic material. Upon bacteriophage infection inside the bacteria, the Cas machinery barcodes small phage genome sequences into the bacteria genome to counter-attack using CRISPR-Cas9 to cleave foreign genetic material [39]. One of their key features is the 'sequence specific targeting,' distinguishing both pathogenic and commensal bacteria. CRISPR-Cas9 system eliminates bacterial virulence factors carried on virulence plasmids and resistance determinants in commensal bacteria [40]. It can enhance the cytotoxicity of the resistant cells due to the nuclease activity of Cas9 [41].

Genome Engineering in Agriculture

Genome engineering can be widely used in agriculture for several applications (Fig. **5**), such as understanding the gene function or altering certain characteristics for a specific function [42]. Genome editing can generate single nucleotide polymorphisms (SNP) [42]. SNP is when one nucleotide can be altered to another nucleotide without the total size of the genome being changed. SNP can even attenuate gene or protein function. When the nucleotide changes, the codon that codes for amino acids also differs, which leads to missense mutations. Indels can be introduced by genome engineering to alter the expression of a specific target to study the phenotype [42]. For example, Os11N3 activation causes sucrose export from plant cells and helps pathogen growth. When indels are brought into the binding site, Os11N3 is not induced, thus inhibiting pathogenicity [43].

Genome editing can create genetically modified (GM) crops with specifically required characteristics in agriculture. Transgenes produce GM crops into elite crop varieties [44]. Disease resistance, high crop yield, resistance to abiotic stress (Fig. **5**), and plant pathogens are some of the characteristics that can be introduced.

Some prominent examples are CRISPR/Cas9 technology which is used to generate Taedr1 wheat plant by modifying 3 homeologs of EDR1 simultaneously. These generated plants were resistant to powdery mildew [45]. OSERF922 mutagenesis results in bacterial blight resistance characteristics. Engineered mutations in SlaGAMOUS-LIKE 6 (SlAGL 6) and SELF – PRUNING 5G (SP5G) by Crispr-Cas caused parthenocarpic phenotype and rapid flowering in tomatoes, respectively [46, 47].

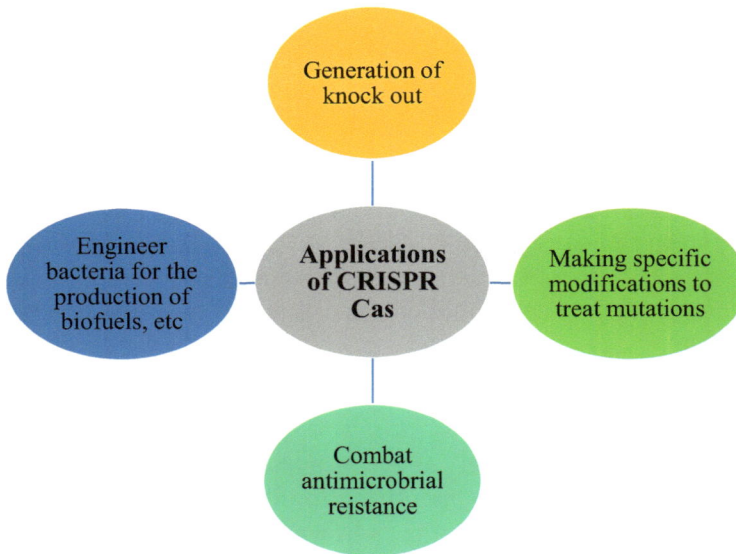

Fig. (5). CRISPR-Cas has various applications, including the generation of knockout, making specific modifications, combatting antimicrobial resistance, and is used in multiple fields like genetics, molecular biology, and agriculture.

CRISPR technology has been utilized to develop crops with improved quality and productivity. Cas12a also has been used in agriculture. LbCas12a is used to create a loss of function alleles of OsEPFL9 that regulates stomatal density, increasing efficiency eight-fold in T2 generation plants [48]. It also aids in site-directed mutagenesis and is a tremendous advantage for targeted gene integration. Thus, genome editing has a significant influence on agriculture (Fig. **5**), and the production era of genome-edited crops is near.

Genome Engineering in Industrial Biotechnology

Microorganisms are used in industrial processes by producing secondary metabolites from low feedstock that can curb environmental pollution. CRISPR – Cas9 system can be used to alter the genetic mechanisms of the bacteria and thereby increase the efficiency of the microorganisms [49]. This genome editing process can predict and modulate virus resistance, transfer nucleic acid to the host, and limit the spread of mobile genetic elements such as transposons. *Clostridium saccharoperbutylacetonicum* N1- 4 is a butanol producing strain, and CRISPR-Cas increased the butanol production from 20% to 75% by choosing the optimized promoter Pj23119 from *E. coli* for gRNA expression [50]. Bulk chemicals can also be produced by introducing CRISPR – Cas9 editing systems in bacteria. Some of the best examples are *Synechoccus elongatus* for succinate production, *Clostridium acetobutylicum* for isopropanol – butanol – ethanol, *and Corynebacterium acetobutylicum* for GABA [51, 52].

This genome engineering concept is widely used in yeast strains as they are actively involved in the production of biopharmaceuticals, chemicals, renewable fuels, *etc*. Herein we discuss recent advances in understanding the diverse mechanisms by which Cas proteins respond to foreign nucleic acids and how these systems have been harnessed for precision genome manipulation in a wide array of organisms [49]. *Scheffersomyces stipites* are generally used for the production of xylose, and the exogenous genes coding for xylose reductase, xylitol dehydrogenase, and xylokinase were integrated into the loci of PHO13 and ALD6 in *S. cerevisiae* by using CRISPR – Cas9 system, and this could be used in large scale fermentation [53].

FUTURE PERSPECTIVES OF CRISPR-Cas9

Allele-specific gene editing prevents deafness in model autosomal dominant progressive hearing loss, a case study: In 2018, a group of scientists from Harvard Medical School experimented with a mouse model for progressive hearing loss (DFNA 36) named Beethoven mice, using CRISPR-Cas9 genome editing to cure the hearing loss in these mice. Beethoven mice are a model organism for studying progressive hearing loss classified as DFNA 36. These mice are named after the famous classical composer Ludwig van Beethoven as it is believed that his progressive hearing loss was also the same disease that is classified as DFNA 36 [54]. DNA 36 is caused by an SNP occurring in the TMC-1 gene, which encodes the TMC1 (Transmembrane channel-like protein1) protein. TMC1 is an essential protein that regulates Ca^{2+} and K^+ ions in the cochlear hair cells. This is crucial for auditory operation, as the cochlea is integral to the inner ear. Proper functioning of the TMC1 protein keeps the hair cells fluid and flaccid, thus allowing the cochlea to receive auditory impulses [55]. For DFNA 36, there is a single nucleotide polymorphism in the TMC1 gene, resulting in faulty copies of TMC1 proteins. This causes the cochlea's hair cells to stiffen, which in turn causes the cochlea to become unable to receive auditory impulses and connect them to the auditory nerve. Beethoven mice have at least one defective copy of this TMC1 gene and become utterly deaf by six months out of a life span averaging nearly 2-3 years [56].

LIMITATION OF CRISPR-Cas

One of the significant limitations of CRISPR is the high frequency of off-target effects. Many unwanted mutations have been observed, and the frequency is above 50%. Scientists are working to reduce the off-target effects by engineering variants of Cas9 and constructing sequence-specific gRNA. The requirement of a

PAM sequence to locate the target sequence is also considered a limitation [57]. Double-stranded breaks trigger apoptosis so that gene editing could be limited to p53 suppressed cells. Thus, immunotoxicity is a limitation taken into consideration [58]. As part of repair mechanisms, HDR can be used for the desired edit but has lower efficiency compared to NHEJ, and thus NHEJ inhibiting enzymes are used to enhance the efficiency of HDR. CRISPR is truly an asset for modern medicine, keeping aside these limitations and some ethical considerations.

CONCLUSION

CRISPR can be a handy tool for editing genes. In the future, the potential of CRISPR-Cas is high. Applying the principle of CRISPR in correcting genetic defects and treating genetic disorders has become widespread. This method is also exploited for commercial purposes to increase the yield of the product in agriculture. Researchers are focussing on enhancing the expression of this method to make it more effective. Hence, CRISPR has established the foundation of treatment and genome editing accurately in this genomic era. It has not yet been approved for usage on humans for clinical trials. We will also likely see approved clinical trials and therapies on adult humans. Each application has its benefits, risks, regulatory frameworks, ethical issues, and societal implications, as with other medical advances. Genome editing has been under great scrutiny. While genome editing has potential applications in agriculture and nonhuman animals, much of the debate is focused on human applications. At this moment, it makes it non-consensual and therefore problematic. The majority of the ethics committees of the world agree that it is irresponsible to use CRISPR on embryos. The future elucidations for effective and precise gene alteration will be found in as-of-yet unknown corners of the vast natural variety of nature.

ACKNOWLEDGMENTS

The authors would like to express their gratitude for the infrastructure and research facilities provided by the Department of Genomic Science, Central University of Kerala and Narayana Medical College.

REFERENCES

[1] Available from: https://www.britannica.com/science/moneran

[2] Maghsoodi A, Chatterjee A, Andricioaei I, Perkins NC. How the phage T4 injection machinery works including energetics, forces, and dynamic pathway. Proc Natl Acad Sci 2019; 116(50): 25097-105.
[http://dx.doi.org/10.1073/pnas.1909298116] [PMID: 31767752]

[3] Barrangou R. The roles of CRISPR–Cas systems in adaptive immunity and beyond. Curr Opin Immunol 2015; 32: 36-41.
[http://dx.doi.org/10.1016/j.coi.2014.12.008] [PMID: 25574773]

[4]   Lau CH. Applications of CRISPR-Cas in bioengineering, biotechnology, and translational research. CRISPR J 2018; 1(6): 379-404.
        [http://dx.doi.org/10.1089/crispr.2018.0026] [PMID: 31021245]

[5]   Rodríguez-Rodríguez DR, Ramírez-Solís R, Garza-Elizondo MA, Garza-Rodríguez ML, Barrera-Saldaña HA. Genome editing: A perspective on the application of CRISPR/Cas9 to study human diseases (Review). Int J Mol Med 2019; 43(4): 1559-74.
        [http://dx.doi.org/10.3892/ijmm.2019.4112] [PMID: 30816503]

[6]   Bandyopadhyay A, Kancharla N, Javalkote VS, Dasgupta S, Brutnell TP. Crispr-cas12a (Cpf1): A versatile tool in the plant genome editing tool box for agricultural advancement. Front Plant Sci 2020; 11: 584151.
        [http://dx.doi.org/10.3389/fpls.2020.584151] [PMID: 33214794]

[7]   Bak RO, Gomez-Ospina N, Porteus MH. Gene editing on center stage. Trends Genet 2018; 34(8): 600-11.
        [http://dx.doi.org/10.1016/j.tig.2018.05.004] [PMID: 29908711]

[8]   Prize N. Emmanuelle charpentier, nobel prize in chemistry 2020: her journey to the nobel prize. 2022. Available from: https://www.youtube.com/watch?v=xirNOTt5dK4

[9]   Ledford H. Super-precise CRISPR tool enhanced by enzyme engineering. Nature 2020.
        [http://dx.doi.org/10.1038/d41586-020-00340-w] [PMID: 33558744]

[10]   Barrangou R, Marraffini LA. CRISPR-Cas systems: Prokaryotes upgrade to adaptive immunity. Mol Cell 2014; 54(2): 234-44.
        [http://dx.doi.org/10.1016/j.molcel.2014.03.011] [PMID: 24766887]

[11]   Aliyari R, Ding SW. RNA-based viral immunity initiated by the Dicer family of host immune receptors. Immunol Rev 2009; 227(1): 176-88.
        [http://dx.doi.org/10.1111/j.1600-065X.2008.00722.x] [PMID: 19120484]

[12]   Dugar G, Herbig A, Förstner KU, *et al.* High-resolution transcriptome maps reveal strain-specific regulatory features of multiple *Campylobacter jejuni* isolates. PLoS Genet 2013; 9(5): e1003495.
        [http://dx.doi.org/10.1371/journal.pgen.1003495] [PMID: 23696746]

[13]   Hatoum-Aslan A, Maniv I, Marraffini LA. Mature clustered, regularly interspaced, short palindromic repeats RNA (crRNA) length is measured by a ruler mechanism anchored at the precursor processing site. Proc Natl Acad Sci 2011; 108(52): 21218-22.
        [http://dx.doi.org/10.1073/pnas.1112832108] [PMID: 22160698]

[14]   Yosef I, Goren MG, Qimron U. Proteins and DNA elements essential for the CRISPR adaptation process in *Escherichia coli.* Nucleic Acids Res 2012; 40(12): 5569-76.
        [http://dx.doi.org/10.1093/nar/gks216] [PMID: 22402487]

[15]   Swarts DC, Mosterd C, Van Passel MWJ, Brouns SJJ. CRISPR interference directs strand specific spacer acquisition. PLoS One 2012; 7(4): e35888.
        [http://dx.doi.org/10.1371/journal.pone.0035888] [PMID: 22558257]

[16]   Wiedenheft B, Sternberg SH, Doudna JA. RNA-guided genetic silencing systems in bacteria and archaea. Nature 2012; 482(7385): 331-8.
        [http://dx.doi.org/10.1038/nature10886] [PMID: 22337052]

[17]   Shah SA, Hansen NR, Garrett RA. Distribution of CRISPR spacer matches in viruses and plasmids of crenarchaeal acidothermophiles and implications for their inhibitory mechanism. Biochem Soc Trans 2009; 37(1): 23-8.
        [http://dx.doi.org/10.1042/BST0370023] [PMID: 19143596]

[18]   Kunin V, Sorek R, Hugenholtz P. Evolutionary conservation of sequence and secondary structures in CRISPR repeats. Genome Biol 2007; 8(4): R61.
        [http://dx.doi.org/10.1186/gb-2007-8-4-r61] [PMID: 17442114]

[19]   Hille F, Richter H, Wong SP, Bratovič M, Ressel S, Charpentier E. The biology of CRISPR-Cas: Backward and forward. Cell 2018; 172(6): 1239-59.
[http://dx.doi.org/10.1016/j.cell.2017.11.032] [PMID: 29522745]

[20]   Deltcheva E, Chylinski K, Sharma CM, *et al.* CRISPR RNA maturation by trans-encoded small RNA and host factor RNase III. Nature 2011; 471(7340): 602-7.
[http://dx.doi.org/10.1038/nature09886] [PMID: 21455174]

[21]   Jore MM, Lundgren M, van Duijn E, *et al.* Structural basis for CRISPR RNA-guided DNA recognition by Cascade. Nat Struct Mol Biol 2011; 18(5): 529-36.
[http://dx.doi.org/10.1038/nsmb.2019] [PMID: 21460843]

[22]   Blum B, Bakalara N, Simpson L. A model for RNA editing in kinetoplastid mitochondria: RNA molecules transcribed from maxicircle DNA provide the edited information. Cell 1990; 60(2): 189-98.
[http://dx.doi.org/10.1016/0092-8674(90)90735-W] [PMID: 1688737]

[23]   Connell GJ, Byrne EM, Simpson L. Guide RNA-independent and Guide RNA-dependent Uridine Insertion into Cytochrome b mRNA in a Mitochondrial Lysate from *Leishmania tarentolae.* J Biol Chem 1997; 272(7): 4212-8.
[http://dx.doi.org/10.1074/jbc.272.7.4212] [PMID: 9020135]

[24]   Maslov DA. Complete set of mitochondrial pan-edited mRNAs in *Leishmania mexicana* amazonensis LV78. Mol Biochem Parasitol 2010; 173(2): 107-14.
[http://dx.doi.org/10.1016/j.molbiopara.2010.05.013] [PMID: 20546801]

[25]   Karvelis T, Gasiunas G, Miksys A, Barrangou R, Horvath P, Siksnys V. crRNA and tracrRNA guide Cas9-mediated DNA interference in *Streptococcus thermophilus.* RNA Biol 2013; 10(5): 841-51.
[http://dx.doi.org/10.4161/rna.24203] [PMID: 23535272]

[26]   Fukuda M, Umeno H, Nose K, Nishitarumizu A, Noguchi R, Nakagawa H. Construction of a guide-RNA for site-directed RNA mutagenesis utilising intracellular A-to-I RNA editing. Sci Rep 2017; 7(1): 41478.
[http://dx.doi.org/10.1038/srep41478] [PMID: 28148949]

[27]   Makarova KS, Wolf YI, Alkhnbashi OS, *et al.* An updated evolutionary classification of CRISPR–Cas systems. Nat Rev Microbiol 2015; 13(11): 722-36.
[http://dx.doi.org/10.1038/nrmicro3569] [PMID: 26411297]

[28]   Miyaoka Y, Berman JR, Cooper SB, *et al.* Systematic quantification of HDR and NHEJ reveals effects of locus, nuclease, and cell type on genome-editing. Sci Rep 2016; 6(1): 23549.
[http://dx.doi.org/10.1038/srep23549] [PMID: 27030102]

[29]   Wu Y, Liang D, Wang Y, *et al.* Correction of a genetic disease in mouse *via* use of CRISPR-Cas9. Cell Stem Cell 2013; 13(6): 659-62.
[http://dx.doi.org/10.1016/j.stem.2013.10.016] [PMID: 24315440]

[30]   Wright AV, Nuñez JK, Doudna JA. Biology and applications of crispr systems: Harnessing nature's toolbox for genome engineering. Cell 2016; 164(1-2): 29-44.
[http://dx.doi.org/10.1016/j.cell.2015.12.035] [PMID: 26771484]

[31]   Bakondi B, Lv W, Lu B, *et al. In vivo* CRISPR/cas9 gene editing corrects retinal dystrophy in the s334ter-3 rat model of autosomal dominant retinitis pigmentosa. Mol Ther 2016; 24(3): 556-63.
[http://dx.doi.org/10.1038/mt.2015.220] [PMID: 26666451]

[32]   Westra ER, Dowling AJ, Broniewski JM, Van Houte S. Evolution and ecology of CRISPR. Annu Rev Ecol Evol Syst 2016; 47(1): 307-31.
[http://dx.doi.org/10.1146/annurev-ecolsys-121415-032428]

[33]   Azangou-Khyavy M, Ghasemi M, Khanali J, *et al.* Crispr/cas: From tumor gene editing to t cell-based immunotherapy of cancer. Front Immunol 2020; 11: 2062.
[http://dx.doi.org/10.3389/fimmu.2020.02062] [PMID: 33117331]

[34]   Ishibashi A, Saga K, Hisatomi Y, Li Y, Kaneda Y, Nimura K. A simple method using CRISPR-Cas9 to knock-out genes in murine cancerous cell lines. Sci Rep 2020; 10(1): 22345.
[http://dx.doi.org/10.1038/s41598-020-79303-0] [PMID: 33339985]

[35]   Peng J, Wang Y, Jiang J, *et al.* Production of human albumin in pigs through crispr/cas9-mediated knockin of human cdna into swine albumin locus in the zygotes. Sci Rep 2015; 5(1): 16705.
[http://dx.doi.org/10.1038/srep16705] [PMID: 26560187]

[36]   Gasiunas G, Barrangou R, Horvath P, Siksnys V. Cas9–crRNA ribonucleoprotein complex mediates specific DNA cleavage for adaptive immunity in bacteria. Proc Natl Acad Sci 2012; 109(39): E2579-86.
[http://dx.doi.org/10.1073/pnas.1208507109] [PMID: 22949671]

[37]   Ran FA, Hsu PD, Lin CY, *et al.* Double nicking by RNA-guided CRISPR-Cas9 for enhanced genome editing specificity. Cell 2013; 154(6): 1380-9.
[http://dx.doi.org/10.1016/j.cell.2013.08.021] [PMID: 23992846]

[38]   Ye L, Wang J, Tan Y, *et al.* Genome editing using CRISPR-Cas9 to create the HPFH genotype in HSPCs: An approach for treating sickle cell disease and β-thalassemia. Proc Natl Acad Sci 2016; 113(38): 10661-5.
[http://dx.doi.org/10.1073/pnas.1612075113] [PMID: 27601644]

[39]   Barrangou R, Fremaux C, Deveau H, *et al.* CRISPR provides acquired resistance against viruses in prokaryotes. Science 2007; 315(5819): 1709-12.
[http://dx.doi.org/10.1126/science.1138140] [PMID: 17379808]

[40]   Law BJC, Zhuo Y, Winn M, *et al.* A vitamin K-dependent carboxylase orthologue is involved in antibiotic biosynthesis. Nat Catal 2018; 1(12): 977-84.
[http://dx.doi.org/10.1038/s41929-018-0178-2]

[41]   Aslam B, Rasool M, Idris A, *et al.* CRISPR-Cas system: A potential alternative tool to cope antibiotic resistance. Antimicrob Resist Infect Control 2020; 9(1): 131.
[http://dx.doi.org/10.1186/s13756-020-00795-6] [PMID: 32778162]

[42]   Baltes N J, Gil-Humanes J, Voytas D F. Genome engineering and agriculture: Opportunities and challenges. ProgMol Biol Transl Sci 2017; 149: 1-26.
[http://dx.doi.org/10.1016/bs.pmbts.2017.03.011]

[43]   Li T, Liu B, Spalding MH, Weeks DP, Yang B. High-efficiency TALEN-based gene editing produces disease-resistant rice. Nat Biotechnol 2012; 30(5): 390-2.
[http://dx.doi.org/10.1038/nbt.2199] [PMID: 22565958]

[44]   Zhang Y, Massel K, Godwin ID, Gao C. Applications and potential of genome editing in crop improvement. Genome Biol 2018; 19(1): 210.
[http://dx.doi.org/10.1186/s13059-018-1586-y] [PMID: 30501614]

[45]   Zhang Y, Bai Y, Wu G, *et al.* Simultaneous modification of three homoeologs of *TaEDR1* by genome editing enhances powdery mildew resistance in wheat. Plant J 2017; 91(4): 714-24.
[http://dx.doi.org/10.1111/tpj.13599] [PMID: 28502081]

[46]   Klap C, Yeshayahou E, Bolger AM, *et al.* Tomato facultative parthenocarpy results from Sl *AGAMOUS-LIKE 6* loss of function. Plant Biotechnol J 2017; 15(5): 634-47.
[http://dx.doi.org/10.1111/pbi.12662] [PMID: 27862876]

[47]   Soyk S, Müller NA, Park SJ, *et al.* Variation in the flowering gene SELF PRUNING 5G promotes day-neutrality and early yield in tomato. Nat Genet 2017; 49(1): 162-8.
[http://dx.doi.org/10.1038/ng.3733] [PMID: 27918538]

[48]   Bandyopadhyay A, Kancharla N, Javalkote VS, Dasgupta S, Brutnell TP. Crispr-cas12a (Cpf1): A versatile tool in the plant genome editing tool box for agricultural advancement. Front Plant Sci 2020; 11: 584151.
[http://dx.doi.org/10.3389/fpls.2020.584151] [PMID: 33214794]

[49]   Tyson G W, Banfield J F. Rapidly evolving CRISPRs implicated in acquired resistance of microorganisms to viruses. Environ Microbiol 2007; 10(1): 200-7.
[http://dx.doi.org/10.1111/j.1462-2920.2007.01444.x]

[50]   Wang S, Dong S, Wang P, Tao Y, Wang Y. Genome editing in *Clostridium saccharoperbutylacetonicum* n1-4 with the crispr-cas9 system. Appl Environ Microbiol 2017; 83(10): e00233-17.
[http://dx.doi.org/10.1128/AEM.00233-17] [PMID: 28258147]

[51]   Wasels F, Jean-Marie J, Collas F, López-Contreras AM, Lopes Ferreira N. A two-plasmid inducible CRISPR/Cas9 genome editing tool for *Clostridium acetobutylicum*. J Microbiol Methods 2017; 140: 5-11.
[http://dx.doi.org/10.1016/j.mimet.2017.06.010] [PMID: 28610973]

[52]   Cho JS, Choi KR, Prabowo CPS, *et al.* CRISPR/Cas9-coupled recombineering for metabolic engineering of *Corynebacterium glutamicum*. Metab Eng 2017; 42: 157-67.
[http://dx.doi.org/10.1016/j.ymben.2017.06.010] [PMID: 28649005]

[53]   Tsai CS, Kong II, Lesmana A, *et al.* Rapid and marker-free refactoring of xylose-fermenting yeast strains with Cas9/CRISPR. Biotechnol Bioeng 2015; 112(11): 2406-11.
[http://dx.doi.org/10.1002/bit.25632] [PMID: 25943337]

[54]   Vreugde S, Erven A, Kros CJ, *et al.* Beethoven, a mouse model for dominant, progressive hearing loss DFNA36. Nat Genet 2002; 30(3): 257-8.
[http://dx.doi.org/10.1038/ng848] [PMID: 11850623]

[55]   Nist-Lund CA, Pan B, Patterson A, *et al.* Improved TMC1 gene therapy restores hearing and balance in mice with genetic inner ear disorders. Nat Commun 2019; 10(1): 236.
[http://dx.doi.org/10.1038/s41467-018-08264-w] [PMID: 30670701]

[56]   György B, Nist-Lund C, Pan B, *et al.* Allele-specific gene editing prevents deafness in a model of dominant progressive hearing loss. Nat Med 2019; 25(7): 1123-30.
[http://dx.doi.org/10.1038/s41591-019-0500-9] [PMID: 31270503]

[57]   Zhang F, Wen Y, Guo X. CRISPR/Cas9 for genome editing: Progress, implications and challenges. Hum Mol Genet 2014; 23(R1): R40-6.
[http://dx.doi.org/10.1093/hmg/ddu125] [PMID: 24651067]

[58]   Ryan OW, Skerker JM, Maurer MJ, *et al.* Selection of chromosomal DNA libraries using a multiplex CRISPR system. eLife 2014; 3: e03703.
[http://dx.doi.org/10.7554/eLife.03703] [PMID: 25139909]

CHAPTER 2

Overview and Applications of CRISPR/Cas9 Based Genome Editing in Industrial Microorganisms

Kiran Dindhoria[1,2], Vivek Manyapu[1], Ashif Ali[1], Aman Kumar[1,2] and Rakshak Kumar[1,2,*]

[1] *Biotechnology Division, CSIR-Institute of Himalayan Bioresource Technology Palampur, Himachal Pradesh-176061, India*

[2] *Academy of Scientific and Innovative Research (AcSIR), Ghaziabad-201002, India*

Abstract: CRISPR-Cas technology has reshaped the field of microbiology. It has improved the microbial strains for better industrial and therapeutic utilization. In this chapter, we have tried to provide an overview of this technology with special reference to its associated applications in the various fields of interest. We have discussed the origin, classification, and different genome editing methods of CRISPR-Cas to understand its historical significance and the basic mechanism of action. Further, different applications in the area of agriculture, food industry, biotherapeutics, biofuel, and other valuable product synthesis were also explained to highlight the advancement of this system in industrial microbes. We have also tried to review some of the limitations offered by CRISPR and insights into its future perspective.

Keywords: Agriculture, Biofuel, Biotherapeutics, CRISPR/Cas9, Food industry, Genome editing, Industrial applications, Microorganisms.

INTRODUCTION

Microorganisms produce many important enzymes and metabolites. They are the source of a variety of industrially valuable products having applications in areas like food, therapeutics, and agronomy. Over the years, several strategies have been applied to improve the industrially important attributes of these microbes for their effective exploitation. With the emergence of genome editing tools such as Zinc finger nucleases (ZFNs) and Transcription-activator effector nucleases (TALENs), the improvement of bacterial strains appeared possible [1]. But despite their good potential, genetic manipulation is hard to achieve in most cases. The main limitations were the cumbersome process of creating novel nuclease

---

[*] **Corresponding author Rakshak Kumar:** Biotechnology Division, CSIR-Institute of Himalayan Bioresource Technology Palampur, Himachal Pradesh-176061, India; & Academy of Scientific and Innovative Research (AcSIR), Ghaziabad-201002, India; Tel: +91 1894 233339; (Ext. 441); Fax: +91 1894 230433; E-mails: rakshak@ihbt.res.in and rakshakacharya@gmail.com

Prakash M. Halami & Aravind Sundararaman (Eds.)

pairs for each target site and their inability to target multiple sites at a time. However, a recently emerged genome editing tool, CRISPR-Cas9, can prove to be very important in the exploration and understanding of the genetic basis for physiochemical and metabolic traits in microorganisms. It is a simple, adaptive, and fast technique that has attracted a lot of scientific interest in the field of genome editing. It can easily be used for targeting multiple genes at a single point of time. It has enabled the rapid genome engineering of bacterial strains, transforming them into better cell factories for the production of value-added products, for instance, *Escherichia coli* [2]. Furthermore, it may be utilized to knock out/in genes and alter somatic genes by genome manipulation even in the germline of species. This method has also been utilized to target, activate, and repress specific genes of interest using specific transcription factors. In 2012, researchers from two independent laboratories suggested that the CRISPR-Cas systems possessing biological functions may be created, which cut off individual target DNA sequences allowing scientists to utilize this tool for genomic manipulations [3]. It has also been employed for checking the possible genome editing in mammals. Cong and associates efficaciously knocked-out numerous genes in each human and mouse cell lines by developing CRISPR/Cas systems with the use of *S. pyogenes* Cas9 (SpCas9) [4]. Similarly, Mali and associates extensively utilized Cas9 to purposefully knock-out genes in numerous human mobileular strains [5].

In this chapter, the CRISPR-Cas system is discussed in detail with some of its applications in improving the different industrially important bacterial strains. Various technical aspects and different types of CRISPR-Cas methods have also been described. We have also tried to explain major challenges and future aspects of the technique.

ORIGIN OF CRISPR/Cas9 SYSTEM

CRISPR is a large family of short palindromic repeat sequences found in a wide range of prokaryotes, including bacteria and archaea. The discovery of CRISPR happened accidentally in 1987 when Ishino *et al.*, were working on the *iap* gene in *E. coli*, which encodes alkaline phosphatase [6], and they identified a set of re-occurring DNA sequences in the bacterial genome that differed significantly from other sequences. Later, comparable CRISPR genes from different bacteria and archaea were also cloned [7 - 10]. However, for over a decade, scientists did not understand the function of these unusual repeating sequences and just assumed that they were unique sequences across different bacterial species. But in 1995, Mojica and associates created a plasmid containing the fragments of CRISPR sequences which were used to transform halophilic archaea *Haloferax volcanii*; and found that extra copies of CRISPR sequences caused alteration in the genome distribution of *H. volcanii* [11]. This, for the first time, revealed incompatibility between external plasmid CRISPR and archaea. However, their actual feature became uncertain till the scientists located CRISPR-related protein (*Cas*) genes and their capabilities in bacterial protection mechanisms. Later, it was discovered that these recurrent DNA sequences might be important components of the "Clustered Regularly Interspaced Short Palindromic Repeats" (CRISPR) family of repetitive DNA sequences [12]. In bacteria, short segments of spacer DNA generated accompany short repeats of DNA sequences from prior exposure to a bacteriophage or plasmid in these CRISPR systems. CRISPR repetitions were discovered to be linked to nucleases or helicases involved in the cleavage or unwinding of particular DNA regions. The primary function of the CRISPR/Cas systems was to protect bacteria from bacteriophage or plasmid invasion. When the system is re-exposed to the same bacteriophage or plasmid, the CRISPR/Cas recognizes it based on its transcribed RNA sequences, and a Cas nuclease is instructed to break the DNA. Cas9 was discovered to be a nuclease with the ability to chop DNA at two active cleaving sites, one for each DNA strand, and was isolated from the bacteria *Streptococcus pyogenes*. A single Cas9 protein may be reused to target and cleave specific locations on bacterial DNA.

CRISPR-Cas Systems Classification

CRISPR-Cas tool is generally categorized into two groups: class 1, including multi-subunit effectors, and class 2, including single protein effectors. Based on their characteristic proteins, these two classes are further categorized into six different types. Three of these have received the most attention: type I, type II, and type III (Fig. **1**), possessing signature proteins Cas3, Cas9, and Cas10, respectively. Most of the CRISPR-Cas systems contain Cas1 and Cas2 proteins, responsible for the integration of the spacer during the adaptation step, in addition to their respective signature proteins [13]. The presence of a protospacer adjacent motif (PAM), corresponding to short conserved sequences, is all that is required for type I and type II systems to target DNA. Furthermore, for the occurrence of DNA interference (DNAi) in all three kinds of CRISPR-Cas systems, a seed sequence with 8–10 base pairs at the 3' end of the guide RNA is required. Type II CRISPR-Cas is the most common and extensively utilized as it relies on a single Cas9 nuclease protein for DNA-induced gene silencing (DNAi). Cas9 is made up of several domains and works in alliance with short RNAs, including mature CRISPR RNA (crRNA) and a trans-acting RNA (tracrRNA) [14]. The HNH system type is also seen in type II systems (also called as Nmeni subtype, for *Neisseria meningitidis* serogroup A str. Z2491). In this system, in addition to the ubiquitous Cas1 and Cas2, the single and extremely big Cas9 protein appears to be ideal for producing CRISPR RNA (crRNA) and slicing the target DNA. Cas9 is made up of two nuclease domains, which are arranged in the order: the RuvC-like nuclease domain is located at the amino terminus, whereas the HNH (or McrAlike) nuclease domain is located in the centre of the protein. The PAM interacting domain (PI domain), which recognizes the presence of the PAM, is extremely close to Cas9's C-terminal end [13].

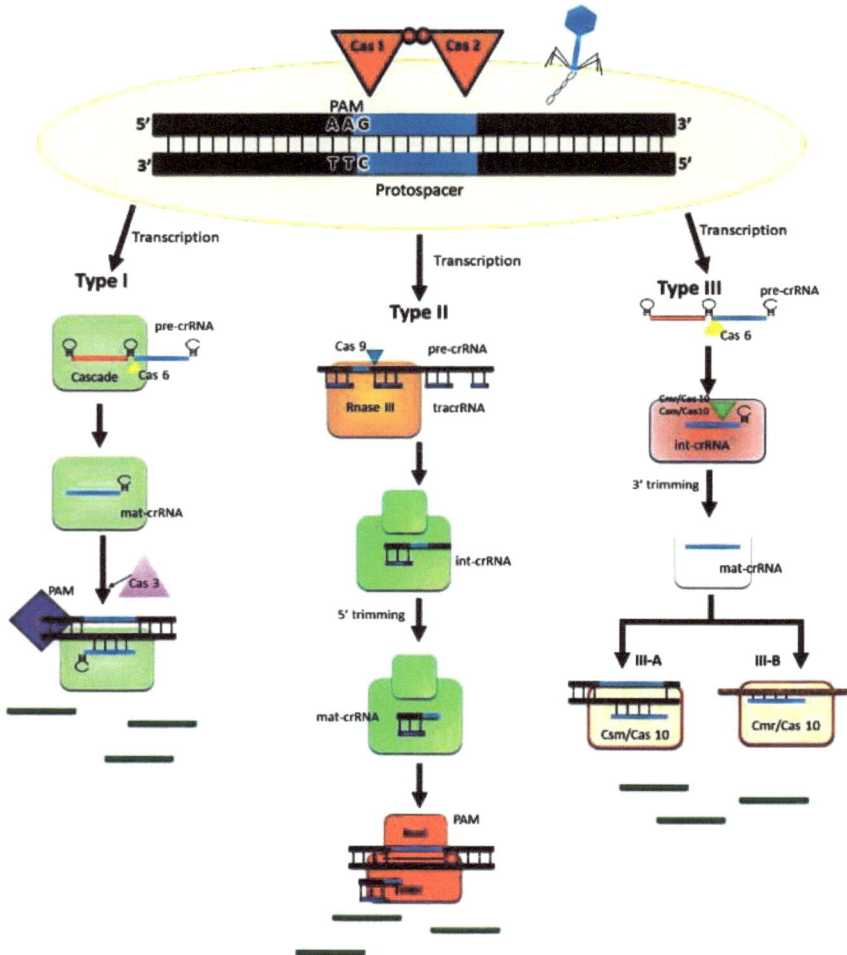

Fig. (1). Mechanisms of CRISPR immunity in all three types of CRISPR-Cas systems. When any bacteriophage invades a bacterial cell, the *Cas* genes present on the CRISPR locus of the bacterial genome get activated and recognize the protospacer adjacent motif (PAM) region of the bacteriophage genome. The *Cas* genes cleave the bacteriophage genome into small fragments. Later one of the fragments gets incorporated into the CRISPR array. The CRISPR array is then transcribed into a long precursor CRISPR RNA (pre-crRNA), which is further processed by Cas6 in type I and III systems. In the case of type II CRISPR-Cas system, trans-activating CRISPR RNA (tracrRNA), RNase III, and Cas9 are involved in the maturation of crRNA (Modified from [92]).

As CRISPR-Cas in bacteria form an RNA-guided adaptive immune system that protects them against invading genetic elements, CRISPR-Cas is a virus-defeating acquired immune defence system developed by prokaryotes. It can also be called a naturally occurring tool for genome editing in bacteria. The various molecular mechanisms of acquisition, expression, and interference, all play a crucial role in CRISPR–Cas mediated immunity [15]. To induce interference, CRISPR RNAs direct Cas proteins to recognize and cleave target DNA complementary to the spacer in a sequence-specific manner. Type I, type II, and type III CRISPR–Cas systems encode universal *Cas*1 and *Cas*2 genes and are classified as Type I, Type II, or Type III based on signature genes that contribute to the different methods by which each system confers interference [16]. CRISPR-Cas system is a large family; in keeping with a current study, it may be divided into two classes, sis types and thirty-three subtypes primarily based totally on numerous criteria, along with collection similarity, awesome functions of the additives, and phylogenetic analysis [16]. But with the passing time, if new CRISPR/Cas are being discovered, this family may expand in the future.

CRISPR/Cas based Genome Editing Methods

Gene Knock-out

Gene knock-out is generally referred to as the interruption of gene function at specific loci, especially using genetic engineering methods. The conventional gene knock-out method is based on homologous gene recombination. It involves a DNA construct with the desired mutation. It typically uses a drug resistance marker in place of target genes for knock-out purposes. The construct contains approximately 1 kb of homology sequence on either side to that of the target sequence. It is delivered to target cells by electroporation or conjugation, and positive clones with reporter markers are knock-outs [17]. Later, another recombinant-based knock-out method was developed using λ-red recombinase and linear DNA, which has slight differences from the conventional method and demonstrated successful deletion of a DNA segment from *E. coli* [18]. However, recently CRISPR/Cas9 genome editing technique has appeared, which is a more advanced, simple, accurate, and the most efficient technique compared to other gene-editing methods. It can be used for genetic manipulations of microbes for better production of secondary metabolites, enzymes, natural products, and pigments for industrial applications. CRISPR/Cas9 based knock-out toolkit is a CRISPR/Cas9 knock-out mechanism that involves the recognition of the target DNA sequence proximal to protospacer adjacent motif (PAM) sequence (NGG) by small engineered single-stranded guide RNA (sgRNA), which initiate double-strand break (DSB), leading to a blunt-end nick [19]. The repairing of double-strand breaks through the non-homologous end-joining (NHEJ) repair system creates a small deletion (indel) or insertion mutation, therefore, interrupting the target gene. Gene knock-out experiments based on CRISPR/Cas9 rely on a non-homologous end joining (NHEJ) repair system. It can be used in the genetic alteration of industrially important microorganisms, where the overproduction of desired target metabolite is hindered due to the unwanted by-products generated through competing pathways is a big challenge. Therefore, the gene knock-out of major enzymes associated with the competing pathways provides an alternative for metabolic flux guided towards required metabolites [20]. CRISPR/Cas9 mediated gene knock-out is an irreversible way to remove the gene function and ultimately inactivate the interfering competing pathways. For instance, *Tatumella citrea* produces a major industrial component, 2-keto-D-gluconic acid, which is a crucial vitamin C precursor. Jiang *et al.* removed a chromosome in *T. citrea* using the CRISPR/Cas9 tool to delete the genes responsible for unwanted enzyme synthesis pathways, thus improving its quality and yield [21]. In another study, *Clostridium acetobutylicum* ATCC 824 xylose degrading ability was improved by deletion of gene *glcG*, which causes catabolite repression leading to the 24% increment of the ABE production comparably with wild type strain [13]. Similarly, *Bacillus* strains producing various valuable products have been enhanced using a CRISPR-base gene knock-out application. Li and colleagues developed a CRISPR/Cas9-based gene-editing tool in *B. licheniformis*. They developed the method for single-gene knock-out, double gene deletion, and large DNA fragment deletion containing gene clusters using the CRISPR-Cas9 system. Single gene knock-out of an essential *yvmC* gene of red pigment pulcherrimin synthesis pathway was targeted, which was deleted with 100% efficiency with the maximum size of homology arms of 1 kb; therefore, the knock-out efficiency was directly proportional to the homologous arm size [22]. CRISPR/Cas system-based gene knock-out and overexpression have also been employed in *Corynebacterium glutamicum* for overproduction of gamma-aminobutyrate (GABA) for industrial application. GABA production was increased through the overexpression of the *gadB* gene and deletion of *gabT, gabP,* and *Ncgl*1221 genes encoding the GABA transaminase, GABA permease, and L-glutamate transporter using the CRISPR/Cas system [2].

Gene Knock-in

Gene knock-in is defined as the insertion of foreign DNA, including promoter, protein-coding genes, and reporter genes at a particular locus of the chromosome. The conventional method of gene knock-in is very similar to knock-out, which is based on homologous recombination. It depends on the homology-directed repair (HDR) pathway, which inserts a new DNA containing the homologous sequence of either side of cleaved DNA. Therefore, for a high success rate of knock-in, the gene insert must have long homologous (~1Kb) flanking overhangs. Knock-in is more successful in those bacteria, where the HDR repair is dominating over the NHEJ repair mechanism. In general, a double-strand break is created with the help of the gRNA–Cas9 complex at the insertion site of DNA followed by the activation of homology-directed repair (HDR). DNA-repair pathway in the presence of donor template containing the target gene at the inserting site of DNA. The 3' prime end of cleaved DNA intake, and polymerize the entire donor template through cleaved strand repair. Finally, the insertion of the target gene into the specific site is completed with the DSB repair. Nakashima and colleagues used this method to knock-in *E. coli* "T7 RNA polymerase gene" cassette a doxycycline-inducible promoter at the lacZ locus [23]. CRISPR/Cas9-based gene knock-in toolkit is rapid, advanced, efficient, and specific. It can be used in imparting new functions to the host microbes like the activation of the multiple silent biosynthetic gene clusters (BGCs) carried out in *Streptomyces* for the production of various novel metabolites like pH-responsive actinorhodin-related metabolites, type II polyketide, red undecylprodigiosin, and other secondary metabolites for industrial application [24]. For this knock-in, a helper plasmid pCRISPomyces-2 construct was generated having a constitutive bidirectional/unidirectional promoter cassette *kasO*p-P8*/*kasO*p* between two adapter sequences 1, 2. A Protospacer of 20 bp was inserted into the construct target cluster with the help of an assembly. The helper plasmid was inserted with the upstream and downstream homology arms (2-Kb) using the specific restriction enzymes. The final Knock-in construct was transformed into *Streptomyces* spp. *via* conjugation using *E. coli* strains, the successfully transformed colonies were selected with antibiotic resistance. Further knock-in confirmation is done through the specific primer. This study demonstrated that CRISPR/Cas9 knock-in with the constitutive promoter can be used to overexpress important and silent metabolites from BGCs for industrial application. Gene knock-in was also demonstrated in *E. coli* where the sulA gene was overexpressed, leading to the enlargement of a cell for the accumulation of PHA granules [25]. Tao Chen and his group also inserted the β-carotene metabolic pathway into the genome of *E. coli* for modulation of the methylerythritol-phosphate (MEP) pathway and central pathways for enhancing the synthesis of β-carotene [26]. Therefore, the CRISPR based knock-in techniques can be used for developing the bacterial strain for both improved and higher production of valuable products.

Transcriptional Regulation

Transcriptional regulation refers to the process of control of DNA to RNA conversion and eventually the gene expression process. It involves several genetic elements such as promoter, transcription factor, sigma factor, coactivator, corepressor, *etc*. The recently emerged CRISPR-Cas system is an advanced and efficient tool for the regulation of gene expression in bacteria, but still, more understanding and improvement are required for this system. It can be utilized for transcriptional regulation depending on the binding of modified Cas effectors on target DNA without creating any DNA breaks [27]. In simple understanding, the Cas9 nuclease can be converted to deactivated Cas9 (dCas9) by mutating two key residues in its nuclease domain. It can inhibit transcription by sterically interfering with the initiation or elongation of the target gene with a properly chosen sgRNA. The strength of repression is dependent on the nature of the promoter present. In prokaryotes, thousand-fold repression was achieved by dCas9 to the targeted DNA sequence in a promoter or downstream. Similarly, transcriptional activation can be carried out with the help of dCas9, sgRNA, or both to recruit transcription effectors to the DNA. Previously transcriptional regulations in the bacterial system were restricted to gene knock-out and inducible promoter techniques, but with the emergence of the CRISPR-Cas system, new doors are opened in this area. Recently, a transcription activator was constructed in *E. coli* by fusion of RNA polymerase ω subunit to dCas9 [28]. It was found to be moderately active with weak promoters but has very less effect with strong promoters. Similarly, Tong and colleagues showed controlled reversible expression of antibiotic actinorhodin in *Streptomyces coelicolor* [29]. Transcriptional regulation was also achieved in filamentous fungi *Penicillium rubens*. In this process, the nuclease-defective mutant of Cas9 was merged with a highly active tripartite activator VP64-p65-Rta (VPR) to carry out sgRNA directed gene regulation [30]. CRISPR-based interference is also capable of inactivating specific genes without cleaving them off. CRISPR-Cas also allows transcriptional regulation with the help of different epigenetic modifications, such as the repression of the transcription process through gene methylation. Conversely, it can also be used for transcriptional activation [31].

Multiplex Editing

Multiplex genome editing refers to the alterations made at multiple genomic loci simultaneously. It enables applications like multigene knock-outs, chromosomal deletions, translocations, epigenetic modifications, transcriptional activation or repression, *etc.*, for the process of genome engineering. Several groups have developed different approaches for CRISPR-Cas system multiplexing. In general, either multiple gRNAs with a promoter and terminator or multiple gRNAs with a single promotor and terminator are expressed [32]. In a study, the production of many gRNAs from one gene transcript was demonstrated using the tRNA-processing endogenous system. They designed a tRNA-gRNA (PTG) polycistronic genome editing tool for efficient manipulations in microbial systems [33]. Multiplexing is beneficial during editing more than two genes concurrently, or a single gRNA is unable to disrupt the target gene. It can prove to be crucial for genetic manipulations regulating multiple genes in slow-growing bacteria like

Mycobacterium tuberculosis, which has a doubling time of 24h. The slow growth of bacteria provides enough time required for multiple gene editing in these strains [34]. It can also be utilized in improving industrially important bacterial strains. Strain improvement using multiple gene editing saves both time and resources. The combination of several gRNA expression toolkits has enabled successful genome editing in various yeasts such as *Saccharomyces cerevisiae*, *Kluveromyces lactis*, *Ogataea polymorpha*, *Yarrowia lipolytica*, *etc* [35]. Otoupal and the group also constructed the combinatorial library of multiplexed, deactivated CRISPR/Cas9 to control multi-drug resistance in *E. coli* [36]. Upgraded screening systems like CRISPR Enabled Trackable Genome Engineering (CREATE) [37] and CRISPR optimized MAGE Recombineering (crMAGE) [38] systems were also developed by combining Cas9 and other recombineering tools. The currently available tools allow us to edit multiple genes at a time with reduced time, but they still need improvement for the rapid upgradation of industrial strains.

Signalling Pathways

Advancement in synthetic biology has allowed the modifications in microorganisms for the production of valuable crucial in different types of industries. Industrial production by microbes requires high titres, rates, and yields to reduce production costs and maximize profit. Production of valuable ingredients by microorganisms *via* fermentations generally requires suitable substrates, co-factors, and available sources of energy feasible signal transduction pathways. The valued end products produced in the fermentation process are affected majorly by the signal transduction. Hence relaying of signal transduction pathways is important for the biosynthesis of valued products and metabolic engineering. Genome engineering with CRISPR/Cas9 has made it possible to redirect signalling pathways for metabolic engineering. One of the most important limitations in the employment of this approach is the absence of sensors required to process the biological signals. Recently, Liu *et al.* have developed signal conductors based on CRISPR/Cas9 controlling the incorporation of the signalling pathway that regulates the transcription of genes in response to an internal or external stimulus [39]. In this methodology, riboswitches were incorporated in sgRNA, which can detect exact signals. This shows that the bacterial signalling pathway can be reprogrammed using the CRISPR leading to strain improvement. Similarly, synthetic chimeric receptors (dCas9-synRs)· acting in signal transduction were also designed to initiate the targeted signalling inside the cells [40]. Regulation of signalling pathway using the CRISPR-Cas system can also be used in controlling unwanted bacterial biofilm formation, thus reducing the virulence and drug resistance of the pathogenic strains. Karlapudi and his team have created a sgRNA toolkit to control the phenomenon of quorum sensing *Acinetobacter* species [41].

APPLICATIONS OF CRISPR/Cas9 IN DIFFERENT TYPES OF INDUSTRIES

CRISPR-Cas technology has emerged as a potent and ubiquitous tool for genome modification, with far-reaching consequences in the fields of biology and medicine. It is a user-friendly technique that plays an important role in agriculture, biotherapeutics, probiotics, food science, enzymes, and biofuel energy generation. Some of the applications exhibited by CRISPR-Cas are as follows:

Agricultural Industry

With the continuous increase in population and changes in climatic conditions, the availability of food resources is under threat for future generations. To deal with the upcoming scarcity of food worldwide, the improved varieties of crops in terms of productivity and stress resistance should be introduced. CRISPR–Cas technology appeared as a powerful technique in the field of agricultural biotechnology due to its capability of manipulating the plant genome precisely. It has not only been used to develop novel plant varieties with desired traits but also aided in their domestication within a short period and revolutionized their current breeding system. Due to its robustness and high target specificity, it is very helpful in genome editing of different crops generating novel germplasms of required traits, thus making the agriculture system sustainable. It has been observed to be used in the generation of many novel plant varieties with higher yield and good quality products, less herbicide/pesticide requirements, resistance to harmful abiotic & biotic parameters, precise breeding, and accelerated domestication. To increase the cereals' yield, Wang and co-workers manipulated cytokinin homeostasis in an effective way among all the other factors. CRISPR-Cas mediated gene editing of C terminus in *Oryza sativa LOGL5*, which encodes a cytokinin-activating enzyme in rice, resulting in higher grain yield in broad environmental conditions [42]. In the same year, Zhu *et al.* also reviewed some CRISPR-Cas edited genes like *O. sativa GW5*, *O. sativa GW2* and *Triticum aestivum GW2* (for grain weight), *O. sativa PIN5b* (for the size of panicle), *O. sativa GS3* (regulating grain size) to develop plants with higher yields [43]. Along with yield, some other characteristics of the crop also need to improve for better agriculture practices. For example, low amylose content in cereal grains aids in better cooking and promotes good health, with their wide applications in adhesive and textile industries. Twelve inbred lines of waxy maize were developed by disrupting the granule-bound starch synthase 1 (*GBSS1*) (Crucial for the biosynthesis of amylose [44], and rice varieties having low amylose content were created by manipulating the amino acid sequences of *GBSS1* using CRISPR-Cas [45]. These alterations can be used in many different ways to generate high-quality plant varieties for agricultural use.

Moreover, it can be used to manipulate the genetic properties of the microbiome present in the rhizosphere of the plant. We all know that microorganisms like bacteria, fungi, and yeasts are commonly found in the rhizosphere of many crops, where they release different types of secondary metabolites responsible for promoting plant growth and health. Bacteria from diverse genera, *Pseudomonas, Bacillus, Azospirillum, Klebsiella, Rhizobium, Azotobacter, Beijerinckia, Burkholderia, Clostridium, Erwinia, Flavobacterium,etc.*, reported showing a variety of plant growth-promoting activities [46]. CRISPR/Cas9 tool can be used to enhance the growth-promoting and stress tolerance properties of these plant growth-promoting *rhizobacteria* (PGPRs). It is a simple, efficient, and specific tool that can make a larger positive impact on the agricultural industry by increasing the properties of PGPRs. For instance, recently, Yin and group have employed CRISPR/Cas9 to study the various aspects of plant growth-promoting activity of rhizosphere-associated bacterium *Bacillus mycoides* EC18 [47]. They found that double gene knock-out *asbA* and *dhbB* of *Bacillus mycoides* EC18 increases the total amount of chlorophyll present and the plant biomass in *Lolium perenne.* The nitrogen fixation and phosphate solubilizing capacity of rhizobacteria which is very helpful for plants can also be increased using genome editing techniques [48]. Setten *et al.* introduced *nif* genes that encode nitrogenase enzyme to PGPR strain *Pseudomonas protegens* Pf-5. X940 from *Pseudomonas stutzeri* A1501 to enhance the growth of *Arabidopsis* & alfalfa and maize & fescue by increasing ammonium ions and nitrates availability in the soil in response to the *nif* genes [49]. Besides promoting plant growth, PGPRs also act as biocontrol agents. It was found that the plants treated with bacteria *Pseudomonas, Bacillus,* and *Serratia* showed a lower occurrence of pathogen attack [50]. CRISPR/Cas9 tool has also been used in developing disease resistance in plants. Fusarium head blight (FHB), which is a dangerous cereal disease all over the globe, showed reduced disease indices ranging from 40 to 80% on durum wheat upon application of RNA interference (RNAi) Δ*tri6* mutants of *Fusarium culmorum* [51]. Similarly, bacterial blight is a severe disease caused by *Xanthomonas oryzae* pv. *oryzae* threatens the global rice supply. The *SWEET* genes are responsible for disease susceptibility, and a set of bacterial agents can trigger their transcription during infections. Through the use of CRISPR–Cas, researchers have created rice lines with broad-spectrum resistance to *Xanthomonas oryzae* pv. *oryzae* by altering the promoter region of *O. sativa SWEET11, O. sativa SWEET13,* and *O. sativa SWEET14* [52]. In short, the CRISPR-Cas system can be employed to create improved varieties of crops, but more information on disease resistance, molecular plant-pathogen interactions, and crop and pathogen genome data are required to develop better CRISPR based genome editing technologies.

Biofuel and Other Industrial Products

Biofuels are a very important alternative to fossil fuels mainly because of their renewability, easy production, and same properties as their conventional petroleum counterparts. They are cost-effective and environmentally friendly. Biofuel generation is generally carried out with the help of microorganisms through various fermentation processes. However, the natural microbial fermentation process gives lower yields, which raises questions about the adaptation of industrial operations on a broad scale. For this reason, more sophisticated technologies, such as fermentation, metabolic engineering, and synthetic biology should play a key role in enhancing the process of biofuel generation [53]. Recently, CRISPR-Cas technology, which targets a specific section of the genetic code and edits DNA at precise targets in the genome, has revolutionized genetic engineering approaches and shown a major influence on existing bio-refineries. Biofuel production may be manipulated in a variety of ways using this simple yet accurate technology. It can aid in increased biomass hydrolysis by identifying and suppressing competitive pathways, as well as better solvent tolerance and substrate utilization [54]. However, the potential off-target effects may be responsible for impeding the success and safety of the process.

Several recent studies have reported the applications of CRISPR/Cas9 mediated genome editing in microbes as well as substrates for enhanced biofuel production. In a study, this technique was used to mutate the *caffeic acid O-methyltransferase 1 (HvCOMT1)* gene in barley responsible for lignin biosynthesis to make it an efficient lignocellulosic feedstock for biofuel production. Mutation in *HvCOMT1* results in a reduction of total lignin content and syringyl (S)/guaiacyl (G), which significantly enhancing the saccharification and bioethanol production efficiency from lignocellulose [55]. Butanol, which is used in internal combustion engines, is also produced from lignocellulosic feedstock. Butanol synthesis and selectivity were also improved by using CRISPR/Cas9 genome engineering for non-model hyper butanol producing *Clostridium saccharoperbutylacetonicum* N1-4 [56]. Similarly, Shanmugam and co-workers reviewed a genetically modified strain of *E. coli* EMJ50 that can generate biobutanol from glucose by over-expressing the endogenous acetoacetyl-CoA thiolase (*thl*), alcohol dehydrogenase (*adhE2*) from *Clostridium acetobutylicum*, and formate dehydrogenase (*fdh1*) from *Candia boidinii* [53]. Also, *E. coli* EMJ50 was reconstructed by combining formate dehydrogenase (*fdh1*) from *C. boidinii* with CoA-acylating propionaldehyde dehydrogenase (*PduP*) from *S. enteric* and alcohol dehydrogenase (*adhA*) from *L. lactis* to produce 0.82 g/L of butanol with yields of 0.068 g/g of glucose under microaerobic conditions [57]. Some other studies suggested that, as a result of carbon catabolite inhibition, glucose in the feedstock prevents *Clostridium* species from using other sugars; however, this may be overcome by manipulating genes involved in sugar absorption [58]. In another research, SpCRISPR-d-Cas9 was used to target the carbon catabolite repression (CCR) of *C. acetobutylicum* DSM792 and *C. pasteurianum* ATCC6013 through the repression of the kinase/phosphorylase (*hprK*) gene, resulting in the co-utilization of glucose and xylose from lignocellulosic feedstock. As a result of this work, biobutanol synthesis utilizing glycerol, a key by-product of the biodiesel industry, has been brought to light. Besides biofuel generation, CRISPR/Cas9 is also used in the biosynthesis of various other industrially important products, as shown in Table **1**. The recent technological advancements in CRISPR-Cas have made it possible to completely re-program microbial metabolism and biosynthesis. Therefore, researchers may further explore these tools to better understand microbial hosts for allowing them to produce different valuable products on a sustainable basis.

Table 1. Industrial applications of CRISPR editing of different types of bacteria.

S.No.	Bacteria	CRISPR/Cas Based Gene Editing	Industrial Application	References
1.	*Escherichia coli*	Cas9	Adipic acid, uridine, and isopropanol production	[8, 93, 94]
2.	*Escherichia coli*	dCas9-based CRISPRi	Terpenoid, malate, naringenin, malonyl-CoA, and mevalonate production	[7, 95 - 97]
3.	*Streptomyces coelicolor*	Cas9	Secondary metabolite production	[9]
4.	*Clostridium cellulovorans*	dCas9-based CRISPRi	Solvent (ethanol, acetone, and butanol) production	[98]
5.	*Bacillus subtilis*	dCas9-based CRISPRi	N-acetylglucosamine and Hyaluronic acid production	[10, 99]

Food Industry

As we all know that the population of the world is increasing continuously, it is necessary to improve quality and sustainability of the food products to fulfil our future needs. Over the years, several technologies have been used for the better development of different food products, but recent progress on CRISPR/Cas9 in the field of molecular biology and genetics has made this process easier. Fermented foods and beverages are being prepared for thousands of years. The properties like nutritional value, flavor, texture, *etc.*, of the fermented food products, mostly depend on their residential microflora. CRISPR/Cas9 can be used to enhance the robustness of starter cultures by editing the lactic acid bacteria (LAB) present in most fermented foods. The development of efficient starter cultures needs screening of strains with naturally occurring desired properties and improvement of these strains for increased efficiency of their characteristics. Identification and typing of important bacterial strains are very big challenges mainly because of their horizontal gene transfer tendency. Although whole-genome sequencing is considered a golden technique for this purpose due to its high cost and time consumption, it can be substituted with CRISPR-based genotyping. It has been applied effectively in many bacterial strains such as *Lactobacillus casei* [59], *Enterococcus faecalis* [60], *Lactobacillus buchneri* [61], *etc.*, exhibiting wide application of CRISPR-Cas systems in bacterial strain typing. Another important and much-required attribute is resistance against bacteriophages in starter cultures during the preparation of fermented foods. CRISPR tool itself provides adaptive immunity against phages and other invasive elements, as discovered by Ishino and team in the year 1987 [6]. The process of immunization involves the integration of DNA sequences from phages at the CRISPR locus. The transcripts produced subsequently acting as small interfering RNAs which guide the cleavage of target foreign DNA.

The lactic acid bacteria present in most of the naturally fermented food products are also called probiotics. These living bacteria when supplied in appropriate amounts impose a health benefit on the host. These living bacteria, when supplied in appropriate amounts not only promote the overall health of the consumer but also modulate gut microflora. They protect the body from the colonization of pathogenic bacteria and stimulate the immune system of the host. Since probiotics can modulate the gut microbiome of the host, they can be engineered to treat a wide range of diseases, including colorectal cancer, pathogenic infections, inflammatory bowel diseases, and metabolic disorders. For example, *L. lactis* was programmed for the constitutive secretion of anti-inflammatory interleukin-10 (IL-10), which is depleted in Inflammatory Bowel Disease patients. Similarly, *Lactobacillus gasseri* was genetically manipulated for the secretion of glucagon-like peptide-1 (GLP-1) to stimulate the conversion of intestinal epithelial cells to insulin-secreting cells [62]. The enterotoxin produced by enterotoxigenic *E. coli* causes diarrhea in susceptible individuals. So, the genetically modified strain of *E. coli* (CWG308) was engineered to produce chimeric lipopolysaccharide by expressing one of *Neisseria meningitidis'* glycosyltransferase, which may imitate host receptors and bind heat-labile enterotoxins with high avidity to avoid infection [63].

Furthermore, CRISPR/Cas9 can be used to enhance the basic attributes of the probiotic strains to improve their efficiency. However, genetic manipulation of probiotic bacteria needs close attention because of the strict regulations on food-associated products. For instance, the attachment of the probiotics in the gastrointestinal tract is a crucial property. It is a well-known fact that the cell wall of the gram-positive bacteria is rich in teichoic acids. It was observed that teichoic acids were conserved among *Lactobacillus* with a potential role in gut adhesion; thus, targeting such cell wall components may also increase the efficiency of probiotic strains [64]. There are a variety of genome editing methods available for expressing or retaining antigens on the cell surface, including fusing antigens with S-layer proteins (SLAPs) [65]. CRISPR-based genome editing can also be used to increase or impose characteristics like stress tolerance and selective sugar catabolism in probiotics. To increase product shelf life, food processors and customers alike would benefit from LABs that have been modified to outcompete rotting bacteria or contain changed metabolic pathways that result in superior texture and enhanced flavor profiles for goods [66].

Biotherapeutics

Despite decades of research, infectious agents like bacteria, viruses, fungi, and parasites are still responsible for many deaths worldwide. The hospitalization rates are increasing and increasing other risks for multidrug-resistant organisms. The illness caused by infectious agents and metabolic disorders such as cancer, diabetes, aging, *etc.*, is burdening the economy on public health systems all over the globe. However, recent breakthroughs in the genome-editing technique CRISPR/Cas9 have enabled scientists to modify and study molecular mechanisms in both pathogens and host. Recently, a group carried out transcriptional inactivation of infectious genes in *M. tuberculosis* using CRISPR/Cas9 interference (CRISPRi) [34]. It can also be utilized to target the DNA of disease-causing bacteria. Vercoe *et al.* based on the capability of CRISPR/Cas9 in cleaving the target plasmid DNA, demonstrating that it can be used to immunize bacteria against the spread of multidrug-resistant plasmids [67] thus controlling the spread of antibiotic resistance. This tool can also be used as an antimicrobial for killing pathogenic microorganisms. Bikard and the group reported the antimicrobial role of CRISPR/Cas9 targeting the methicillin resistance gene for the eradication of MRSA strains from mixed bacterial populations [68]. It can also be involved in the modulation of host immune response; for example, it has been observed to regulate the process of quorum sensing in bacteria like *Klebsiella pneumoniae, Escherichia coli,* and *Pseudomonas aeruginosa* by modifying the recognition of regulator protein LasR in order to reduce host inflammatory response [69].

CRISPR/Cas9 can be used in the treatment of metabolic disorders too. Diabetes, affecting 425 million people in the world, can be treated through gene deletions or target DNA manipulations [70]. Over the years, several simple, rapid, and low-cost systems were also developed to identify and study the genes involved in causing different types of cancers in humans. A study conducted by Wang and co-workers found that CRISPR engineered adipocytes reduce obesity and metabolic disorders in mice caused due to diet [71]. Recently, CRISPR-based tools have also been shown to play an important role in the rapid and specific diagnosis of COVID-19 [72]. It also has the potential to inhibit viral replication, as demonstrated by the Broad Institute of the Massachusetts Institute of Technology and Harvard University, USA, where the enzyme Cas13 inhibited influenza A virus and vesicular stomatitis virus in cell culture [73]. It is feasible to the inhibition of replication of the human influenza virus (HIV) and possesses the ability to deactivate/reactivate HIV latent infections [74]. On the whole, CRISPR-based genome editing tools provide a wide range of applications in the field of biotherapeutics. It provides new opportunities for pharma industries and should be investigated thoroughly considering different aspects of the therapeutic intended to be developed.

BARRIERS TO GENOME EDITING BY CRISPR/Cas9

The CRISPR/Cas9 genome editing technologies are transforming bacterial genome editing processes. Despite their immense potential for altering bacterial genomes, they nevertheless face considerable challenges. The absence of an effective mechanism for editing DNA in non-model microorganisms, the lack of appropriate procedures for the synthesis and delivery of single guide RNA (sgRNA), the lack of combinatorial approaches with other genetic techniques, and off-target effects are some common roadblocks (*i.e.*, undesirable mutations) in this process.

Absence of Efficient DNA Repair Methods

The application of CRISPR–Cas9 systems evokes double-strand breaks (DSBs) in the targeted DNA. In bacteria, DSBs are generally repaired either by homology-directed repair (HDR) or nonhomologous end-joining (NHEJ). They mostly rely on the homologous recombination pathway, which utilizes an undamaged DNA template for the repair of DSBs. The template used is generally the intact homologous DNA, mainly the sister chromosome obtained upon chromosomal replication occurring during the state of division. HDR can accurately edit the genome in many ways through gene knock-in, gene knock-out, replacement, and point mutations [75]. Alternatively, a minor bacterial class carries out NHEJ to amend DNA breaks in the absence of any DNA template with target genomic locus modification. It requires two simple proteins Ku and Lig D (DNA ligase). The protein Ku binds DNA and protects the ends of DSBs. It recruits Lig D, which seals the ends of DSBs to re-join [76]. NHEJ is ideal for dealing with DSBs since it does not exhibit any sequence requirements, but as the broken DNA ends are damaged and need modification, it tends to cause base insertion or deletion (indel) during the repair. It can also lead to a frameshift mutation, which disrupts the target genes in the absence of a homology repair donor. In some bacterial species such as *Mycobacterium* and *Bacillus*, it protects the bacterial genomes from DSBs and promotes genetic variability [77]. Although it is perfect for managing DSBs and does not require both selective marker and donor DNA template, it's not prevalent among prokaryotes. The lethality caused by Cas9-generated DSBs in bacteria is generally associated with the lack of an NHEJ repair mechanism [78]. However, there are some exceptions to this bacterium *Mycobacterium smegmatis,* which dies due to DSBs besides possessing NHEJ systems [79]. Further, in a study conducted by Tao *et al.* in *Clostridium cellulolyticum* using *Streptococcus pyogenes* CRISPR/Cas9 system [80]. They found a new approach for genome editing in *C. cellulolyticumvia* the employment of Cas9n-mediated single-nick generation and HR (SNHR). This mechanism can escape the lethality caused by DSB and allow genome manipulations in bacteria lacking DSB repair systems.

Low Specificity

CRISPR/Cas9 causes off-target mutations in the host genome by producing non-targeted genomic changes. Off-target mutations are critical in the evaluation of genome manipulation tools. They are undesirable while doing metabolic engineering in bacteria. In an *in vitro* experiment, Jinek *et al.* discovered that the annealing of a guide to target DNA can tolerate up to five mismatches, which is a low specificity and a key worry for genome editing for therapeutical and agronomical purposes [19]. The occurrence of off-target mutations is more often associated with the regions which differ by less than three nucleotides than the original sequence, provided they are adjacent to the protospacer adjacent motif (PAM) sequence. Off-target effects are notoriously difficult to track down, needing an entire genome sequence to eliminate them. Various approaches have been used to identify off-targets for example potential sites can be examined using bioinformatics. The addition of two extra guanine nucleotides to the 5' end of truncated guide RNA (gRNA), truncated within the crRNA-derived sequence or truncated gRNA, truncated within the crRNA-derived sequence is the latest advancement of the CRISPR/Cas9 systems meant to decrease off-target mutations [81]. Researchers have also developed many web tools to improve the design of gRNAs like Cas-Off finder [82], CRISPR tool [83], and DeepCRISPR [84]. Ran *et al.* also employed "paired nickase" to reduce the off-target mutations [85]. This method uses D10A Cas9 and two sgRNAs complementary to the surrounding area on opposite strands of the target site. This method creates DSBs in target DNA but only generates single nickases in untargeted locations thus reducing off-target effects. Although many significant attempts have been made to reduce the occurrence of unwanted mutations, further progress is needed, particularly in the field of therapeutic interventions that require exact transformations.

Lack of Suitable Delivery Methods

Efficient delivery of sgRNA and Cas9 is required for the effective application of CRISPR/Cas9 genome editing tools. It can be delivered through many different methods like sgRNA and Cas9 encoding expression vectors, *in vitro* transcribed Cas9 mRNA and sgRNA, or as an RNP complex of Cas9 recombinant protein and sgRNA [86]. Electroporation, lipofection, nucleofection, and microinjection are some of the most popular delivery methods used but due to their limited efficiency on target delivery, poor editing, and control of Cas9, viral methods have been developed. Viral vectors such as lentivirus, adenovirus, and adeno-associated vectors (AAV) are some examples. Adenoviruses and lentiviruses are both not favoured because they elicit immune responses. Also, lentivirus gets integrated into the host genome, which is also a great risk. Whereas AAVs are preferred for CRISPR cargo delivery because of their unique characteristics, including lower pathogenicity, higher transduction efficiency, non-integrating nature, and serological compatibility [87]. Although AAVs are preferred delivery systems, their capacity is limited to ~4.5 kb. Therefore, different strategies have been employed for the proper delivery of CRISPR/Cas9 cargo: Cas9 and sgRNA packaged in separate vectors, split Cas9 packed in parallel AAVs, shorter sized Cas9 from *Staphylococcus aureus, etc* [86]. Recently, gold nanoparticles covered with donor templates and other CRISPR components were also employed as delivery systems [88]. However, due to lower precision in genome editing, their use has been limited. Moreover, in bacterial strains, systematic multiple gene regulation is important for metabolite production. So, a CRISPR/Cas9 genome editing system with sgRNA directing Cas9 to multiple target DNA for editing, activation, and repression of genes is required. However, such genome editing tools which can alter the genome at multiple sites are still lacking.

Ethical Issues

For many years, extensive research on the advancement of cellular repair mechanisms to manipulate DNA *via* genome editing has been in progress. Through this, we can have the power to correct any unwanted mutation or introduce new functions in the genome of the host cell. However, the rapid development of CRISPR-Cas9 has triggered many ethical concerns in bacteria and other organisms. The crucial point is that the benefits of the technique should be higher than the risks involved. The non-targeted effects of CRISPR-Cas9 should be explored in detail because they may transfer from one generation to another and their effect may increase as the generation progresses, thus posing a threat to the host organism and its environment. Undesirable mutations may cause modifications at other places in the genome leading to further mutations which cell death or cell transformation [89]. Another concern is the transfer of these modified genetic elements to wild-type strains or other species. In that case, there is the risk of transfer of the negative effects to their environment is very high. It would be very challenging to regulate the gene drive train in the natural environment, which may cause species extinction and eventually disturb the balance of an ecosystem. Therefore, proper safety precautions are needed to prevent the dissemination of genetically modified strains that pose a threat to the ecosystem or human health. Also, besides the precise genetic manipulations carried by CRISPR-Cas9, it's very hard to identify the genetically modified organism when it's in the environment of the market. Regulation of such organisms is extremely important, and their consumption should be approved by agencies like the Food and Drug Administration. However, the methods which must be employed for such controls are still uncertain.

FUTURE OUTLOOK

Genome editing is very critical in industrial bacteria for strain improvement. However, very few numbers of species have been explored with CRISPR-Cas9 implementation. The CRISPR-Cas9 systems are not universally present endogenously in all bacteria, so the presence of an endogenous system should serve as a starting point for CRISPR-based tools development. Further, to promote its wide range of applications, extensive screening of natural resources for novel CRISPR-Cas9 is required. For example, *Geobacillus thermodenitrificans,* a thermostable Cas9 orthologue has recently been identified. It was isolated from compost and has been recently developed to be used as Cas-9-based genome editing for thermophilic bacteria [90]. Moreover, additional nucleases and their subtypes can be explored, which can improve the genome editing effectiveness in bacteria. For instance, recently, type V CRISPR nucleases were discovered, offering a wide variety of catalytic activity [91]. Also, in this chapter, we have discussed different barriers involved in the application of the CRISPR-Cas9 system. Strategies should be developed to remove its limitations and improve specificity related to target or intended site editing, DSBs repair mechanisms, proper delivery systems, *etc*. At present, no single approach is effective for editing across all bacterial strains, but several strategies should be used in parallel to achieve the desirables. In-depth studies are required to decide general rules for each strategy used for genome editing in multiple bacterial strains. The insights gained during these studies may provide some predictability to CRISPR-based, tools thus promoting their utilization in bacteria.

In conclusion, the CRISPR-Cas9 genome editing technique has redesigned the standards for genome editing in different organisms including bacteria. It is an extremely versatile and precise genome editing tool. It can also prove to be a safe method for strain improvement of industrially important bacteria. It may increase the bio-production of different types of valuable products, thus increasing the overall yield of these bacteria as compared to the wild-type strains. However, despite many applications provided by the CRISPR-Cas9 tool, several challenges need to be resolved for its better efficiency.

ACKNOWLEDGMENTS

KD is thankful to CSIR, Govt. of India for 'Research Fellowship' Grant CSIR-NET JRF award no: 31/054(0139)/2019-EMR-I / CSIR-NET JRF JUNE 2017. The authors acknowledge the financial support from the CSIR in-house project (MLP 0145, MLP 0201), Science and Engineering Research Board Start-up research grant no. SRG/2019/001071 and DST-TDT project no. DST/TDT/WM/2019/43. This manuscript represents CSIR-IHBT communication no. 5138.

REFERENCES

[1]   Carroll D. Genome engineering with zinc-finger nucleases. Genetics 2011; 188(4): 773-82.
[http://dx.doi.org/10.1534/genetics.111.131433] [PMID: 21828278]

[2]   Cho S, Shin J, Cho BK. Applications of CRISPR/Cas system to bacterial metabolic engineering. Int J Mol Sci 2018; 19(4): 1089.
[http://dx.doi.org/10.3390/ijms19041089] [PMID: 29621180]

[3]   Gasiunas G, Barrangou R, Horvath P, Siksnys V. Cas9-crRNA ribonucleoprotein complex mediates specific DNA cleavage for adaptive immunity in bacteria. Proceedings of the National Academy of Sciences of the United States of America 2012 Sep 25; 109(39): E2579-86. Epub 2012 Sep 4.
[http://dx.doi.org/10.1073/pnas.1208507109] [PMID: 22949671] [PMCID: PMC3465414]

[4]   Cong L, Ran FA, Cox D, *et al.* Multiplex genome engineering using CRISPR/Cas systems. Science 2013; 339(6121): 819-23.
[http://dx.doi.org/10.1126/science.1231143] [PMID: 23287718]

[5]   Mali P, Yang L, Esvelt KM, *et al.* RNA-guided human genome engineering *via* Cas9. Science 2013; 339(6121): 823-6.
[http://dx.doi.org/10.1126/science.1232033] [PMID: 23287722]

[6]   Ishino Y, Shinagawa H, Makino K, Amemura M, Nakata A. Nucleotide sequence of the iap gene, responsible for alkaline phosphatase isozyme conversion in *Escherichia coli*, and identification of the gene product. J Bacteriol 1987; 169(12): 5429-33.
[http://dx.doi.org/10.1128/jb.169.12.5429-5433.1987] [PMID: 3316184]

[7]   Chu LL, Dhakal D, Shin HJ, Jung HJ, Yamaguchi T, Sohng JK. Metabolic engineering of *Escherichia coli* for enhanced production of naringenin 7-sulfate and its biological activities. Front Microbiol 2018; 9: 1671.
[http://dx.doi.org/10.3389/fmicb.2018.01671] [PMID: 30100899]

[8]   Wu H, Li Y, Ma Q, *et al.* Metabolic engineering of *Escherichia coli* for high-yield uridine production. Metab Eng 2018; 49: 248-56.
[http://dx.doi.org/10.1016/j.ymben.2018.09.001] [PMID: 30189293]

[9]   Li L, Wei K, Zheng G, *et al.* CRISPR-Cpf1-assisted multiplex genome editing and transcriptional repression in *Streptomyces*. Appl Environ Microbiol 2018; 84(18): e00827-18.
[http://dx.doi.org/10.1128/AEM.00827-18] [PMID: 29980561]

[10] Westbrook AW, Ren X, Oh J, Moo-Young M, Chou CP. Metabolic engineering to enhance heterologous production of hyaluronic acid in *Bacillus subtilis*. Metab Eng 2018; 47: 401-13.
[http://dx.doi.org/10.1016/j.ymben.2018.04.016] [PMID: 29698777]

[11] Mojica FJM, Ferrer C, Juez G, Rodríguez-Valera F. Long stretches of short tandem repeats are present in the largest replicons of the Archaea *Haloferax mediterranei* and *Haloferax volcanii* and could be involved in replicon partitioning. Mol Microbiol 1995; 17(1): 85-93.
[http://dx.doi.org/10.1111/j.1365-2958.1995.mmi_17010085.x] [PMID: 7476211]

[12] Jansen R, Embden JDA, Gaastra W, Schouls LM. Identification of genes that are associated with DNA repeats in prokaryotes. Mol Microbiol 2002; 43(6): 1565-75.
[http://dx.doi.org/10.1046/j.1365-2958.2002.02839.x] [PMID: 11952905]

[13] Fokum E, Zabed HM, Guo Q, *et al.* Metabolic engineering of bacterial strains using CRISPR/Cas9 systems for biosynthesis of value-added products. Food Biosci 2019; 28: 125-32.
[http://dx.doi.org/10.1016/j.fbio.2019.01.003]

[14] Bhaya D, Davison M, Barrangou R. CRISPR-Cas systems in bacteria and archaea: versatile small RNAs for adaptive defense and regulation. Annu Rev Genet 2011; 45(1): 273-97.
[http://dx.doi.org/10.1146/annurev-genet-110410-132430] [PMID: 22060043]

[15] Barrangou R, Marraffini LA. CRISPR-Cas systems: Prokaryotes upgrade to adaptive immunity. Mol Cell 2014; 54(2): 234-44.
[http://dx.doi.org/10.1016/j.molcel.2014.03.011] [PMID: 24766887]

[16] Makarova KS, Wolf YI, Iranzo J, *et al.* Evolutionary classification of CRISPR–Cas systems: A burst of class 2 and derived variants. Nat Rev Microbiol 2020; 18(2): 67-83.
[http://dx.doi.org/10.1038/s41579-019-0299-x] [PMID: 31857715]

[17] Ortiz-Martín I, Macho AP, Lambersten L, Ramos C, Beuzón CR. Suicide vectors for antibiotic marker exchange and rapid generation of multiple knockout mutants by allelic exchange in Gram-negative bacteria. J Microbiol Methods 2006; 67(3): 395-407.
[http://dx.doi.org/10.1016/j.mimet.2006.04.011] [PMID: 16750581]

[18] Datsenko KA, Wanner BL. One-step inactivation of chromosomal genes in *Escherichia coli* K-12 using PCR products. Proc Natl Acad Sci 2000; 97(12): 6640-5.
[http://dx.doi.org/10.1073/pnas.120163297] [PMID: 10829079]

[19] Jinek M, Chylinski K, Fonfara I, Hauer M, Doudna JA, Charpentier E. A programmable dual-RNA–guided DNA endonuclease in adaptive bacterial immunity. science 2012; 337(6096): 816-21.

[20] Pickens LB, Tang Y, Chooi YH. Metabolic engineering for the production of natural products. Annu Rev Chem Biomol Eng 2011; 2(1): 211-36.
[http://dx.doi.org/10.1146/annurev-chembioeng-061010-114209] [PMID: 22432617]

[21] Jiang Y, Chen B, Duan C, Sun B, Yang J, Yang S. Multigene editing in the *Escherichia coli* genome *via* the CRISPR-Cas9 system. Appl Environ Microbiol 2015; 81(7): 2506-14.
[http://dx.doi.org/10.1128/AEM.04023-14] [PMID: 25636838]

[22] Li K, Cai D, Wang Z, He Z, Chen S. Development of an efficient genome editing tool in *Bacillus licheniformis* using CRISPR-Cas9 nickase. Appl Environ Microbiol 2018; 84(6): e02608-17.
[http://dx.doi.org/10.1128/AEM.02608-17] [PMID: 29330178]

[23] Nakashima N, Tamura T. Gene silencing in *Escherichia coli* using antisense RNAs expressed from doxycycline-inducible vectors. Lett Appl Microbiol 2013; 56(6): 436-42.
[http://dx.doi.org/10.1111/lam.12066] [PMID: 23480057]

[24]   Zhang MM, Wong FT, Wang Y, *et al.* CRISPR–Cas9 strategy for activation of silent *Streptomyces* biosynthetic gene clusters. Nat Chem Biol 2017; 13(6): 607-9.
[http://dx.doi.org/10.1038/nchembio.2341] [PMID: 28398287]

[25]   Wang Y, Wu H, Jiang X, Chen GQ. Engineering *Escherichia coli* for enhanced production of poly(3-hydroxybutyrate-co-4-hydroxybutyrate) in larger cellular space. Metab Eng 2014; 25: 183-93.
[http://dx.doi.org/10.1016/j.ymben.2014.07.010] [PMID: 25088357]

[26]   Li Y, Lin Z, Huang C, *et al.* Metabolic engineering of *Escherichia coli* using CRISPR–Cas9 meditated genome editing. Metab Eng 2015; 31: 13-21.
[http://dx.doi.org/10.1016/j.ymben.2015.06.006] [PMID: 26141150]

[27]   Vigouroux A, Bikard D. CRISPR tools to control gene expression in bacteria. Microbiol Mol Biol Rev 2020; 84(2): e00077-19.
[http://dx.doi.org/10.1128/MMBR.00077-19] [PMID: 32238445]

[28]   Bikard D, Jiang W, Samai P, Hochschild A, Zhang F, Marraffini LA. Programmable repression and activation of bacterial gene expression using an engineered CRISPR-Cas system. Nucleic Acids Res 2013; 41(15): 7429-37.
[http://dx.doi.org/10.1093/nar/gkt520] [PMID: 23761437]

[29]   Tong Y, Charusanti P, Zhang L, Weber T, Lee SY. CRISPR-Cas9 based engineering of actinomycetal genomes. ACS Synth Biol 2015; 4(9): 1020-9.
[http://dx.doi.org/10.1021/acssynbio.5b00038] [PMID: 25806970]

[30]   Mózsik L, Hoekzema M, de Kok NAW, Bovenberg RAL, Nygård Y, Driessen AJM. CRISPR-based transcriptional activation tool for silent genes in filamentous fungi. Sci Rep 2021; 11(1): 1118.
[http://dx.doi.org/10.1038/s41598-020-80864-3] [PMID: 33414495]

[31]   Liao HK, Hatanaka F, Araoka T, *et al. In vivo* target gene activation *via* CRISPR/Cas9-mediated trans-epigenetic modulation. Cell 2017; 171(7): 1495-1507.e15.
[http://dx.doi.org/10.1016/j.cell.2017.10.025] [PMID: 29224783]

[32]   Minkenberg B, Wheatley M, Yang Y. CRISPR/Cas9-enabled multiplex genome editing and its application. Prog Mol Biol Transl Sci 2017; 149: 111-32.
[http://dx.doi.org/10.1016/bs.pmbts.2017.05.003] [PMID: 28712493]

[33]   Xie K, Yang Y. A multiplexed CRISPR/Cas9 editing system based on the endogenous tRNA processing.Plant Genome Editing with CRISPR Systems. Springer 2019; pp. 63-73.
[http://dx.doi.org/10.1007/978-1-4939-8991-1_5]

[34]   Choudhary E, Thakur P, Pareek M, Agarwal N. Gene silencing by CRISPR interference in mycobacteria. Nat Commun 2015; 6(1): 6267.
[http://dx.doi.org/10.1038/ncomms7267] [PMID: 25711368]

[35]   Wang L, Deng A, Zhang Y, *et al.* Efficient CRISPR–Cas9 mediated multiplex genome editing in yeasts. Biotechnol Biofuels 2018; 11(1): 277.
[http://dx.doi.org/10.1186/s13068-018-1271-0] [PMID: 30337956]

[36]   Otoupal PB, Cordell WT, Bachu V, Sitton MJ, Chatterjee A. Multiplexed deactivated CRISPR-Cas9 gene expression perturbations deter bacterial adaptation by inducing negative epistasis. Commun Biol 2018; 1(1): 129.
[http://dx.doi.org/10.1038/s42003-018-0135-2] [PMID: 30272008]

[37]   Garst AD, Bassalo MC, Pines G, *et al.* Genome-wide mapping of mutations at single-nucleotide resolution for protein, metabolic and genome engineering. Nat Biotechnol 2017; 35(1): 48-55.
[http://dx.doi.org/10.1038/nbt.3718] [PMID: 27941803]

[38]   Ronda C, Pedersen LE, Sommer MOA, Nielsen AT. CRMAGE: CRISPR optimized mage recombineering. Sci Rep 2016; 6(1): 19452.
[http://dx.doi.org/10.1038/srep19452] [PMID: 26797514]

[39]   Liu Y, Zhan Y, Chen Z, *et al.* Directing cellular information flow *via* CRISPR signal conductors. Nat
       Methods 2016; 13(11): 938-44.
       [http://dx.doi.org/10.1038/nmeth.3994] [PMID: 27595406]

[40]   Baeumler TA, Ahmed AA, Fulga TA. Engineering synthetic signaling pathways with programmable
       dCas9-based chimeric receptors. Cell Rep 2017; 20(11): 2639-53.
       [http://dx.doi.org/10.1016/j.celrep.2017.08.044] [PMID: 28903044]

[41]   Karlapudi AP, T C V, Tammineedi J, Srirama K, Kanumuri L, Prabhakar Kodali V. *In silico* sgRNA
       tool design for CRISPR control of quorum sensing in *Acinetobacter* species. Genes Dis 2018; 5(2):
       123-9.
       [http://dx.doi.org/10.1016/j.gendis.2018.03.004] [PMID: 30258941]

[42]   Wang C, Wang G, Gao Y, *et al.* A cytokinin-activation enzyme-like gene improves grain yield under
       various field conditions in rice. Plant Mol Biol 2020; 102(4-5): 373-88.
       [http://dx.doi.org/10.1007/s11103-019-00952-5] [PMID: 31872309]

[43]   Zhu H, Li C, Gao C. Applications of CRISPR–Cas in agriculture and plant biotechnology. Nat Rev
       Mol Cell Biol 2020; 21(11): 661-77.
       [http://dx.doi.org/10.1038/s41580-020-00288-9] [PMID: 32973356]

[44]   Gao H, Gadlage MJ, Lafitte HR, *et al.* Superior field performance of waxy corn engineered using
       CRISPR–Cas9. Nat Biotechnol 2020; 38(5): 579-81.
       [http://dx.doi.org/10.1038/s41587-020-0444-0] [PMID: 32152597]

[45]   Xu Y, Lin Q, Li X, *et al.* Fine-tuning the amylose content of rice by precise base editing of the *Wx*
       gene. Plant Biotechnol J 2021; 19(1): 11-3.
       [http://dx.doi.org/10.1111/pbi.13433] [PMID: 32558105]

[46]   Kaymak HC. Potential of PGPR in agricultural innovations. Plant growth and health 753 promoting
       bacteria 2010; 45-79.
       [http://dx.doi.org/10.1007/978-3-642-13612-2_3]

[47]   Yi Y, Li Z, Song C, Kuipers OP. Exploring plant-microbe interactions of the rhizobacteria *Bacillus
       subtilis* and *Bacillus mycoides* by use of the CRISPR-Cas9 system. Environ Microbiol 2018; 20(12):
       4245-60.
       [http://dx.doi.org/10.1111/1462-2920.14305] [PMID: 30051589]

[48]   Haskett TL, Tkacz A, Poole PS. Engineering rhizobacteria for sustainable agriculture. ISME J 2021;
       15(4): 949-64.
       [http://dx.doi.org/10.1038/s41396-020-00835-4] [PMID: 33230265]

[49]   Setten L, Soto G, Mozzicafreddo M, *et al.* Engineering *Pseudomonas protegens* Pf-5 for nitrogen
       fixation and its application to improve plant growth under nitrogen-deficient conditions. PLoS One
       2013; 8(5): e63666.
       [http://dx.doi.org/10.1371/journal.pone.0063666] [PMID: 23675499]

[50]   Kloepper JW, Ryu CM, Zhang S. Induced systemic resistance and promotion of plant growth by
       Bacillus spp. Phytopathology 2004; 94(11): 1259-66.
       [http://dx.doi.org/10.1094/PHYTO.2004.94.11.1259] [PMID: 18944464]

[51]   Scherm B, Orrù M, Balmas V, *et al.* Altered trichothecene biosynthesis in TRI6-silenced
       transformants of *Fusarium culmorum* influences the severity of crown and foot rot on durum wheat
       seedlings. Mol Plant Pathol 2011; 12(8): 759-71.
       [http://dx.doi.org/10.1111/j.1364-3703.2011.00709.x] [PMID: 21726376]

[52]   Oliva R, Ji C, Atienza-Grande G, *et al.* Broad-spectrum resistance to bacterial blight in rice using
       genome editing. Nat Biotechnol 2019; 37(11): 1344-50.
       [http://dx.doi.org/10.1038/s41587-019-0267-z] [PMID: 31659337]

[53]   Shanmugam S, Ngo HH, Wu YR. Advanced CRISPR/Cas-based genome editing tools for microbial
       biofuels production: A review. Renew Energy 2020; 149: 1107-19.

[http://dx.doi.org/10.1016/j.renene.2019.10.107]

[54]   Bala A, Singh B. Cellulolytic and xylanolytic enzymes of thermophiles for the production of renewable biofuels. Renew Energy 2019; 136: 1231-44.
[http://dx.doi.org/10.1016/j.renene.2018.09.100]

[55]   Lee JH, Won HJ, Hoang Nguyen Tran P, Lee S, Kim H-Y, Jung JH. Improving lignocellulosic biofuel production by CRISPR/Cas9-mediated lignin modification in barley. Glob Change Biol Bioenergy 2021; 13(4): 742-52.
[http://dx.doi.org/10.1111/gcbb.12808]

[56]   Wang S, Dong S, Wang P, Tao Y, Wang Y. Genome editing in *Clostridium saccharoperbutylacetonicum* N1-4 with the CRISPR-Cas9 system. Appl Environ Microbiol 2017; 83(10): e00233-17.
[http://dx.doi.org/10.1128/AEM.00233-17] [PMID: 28258147]

[57]   Bogorad IW, Chen CT, Theisen MK, *et al.* Building carbon–carbon bonds using a biocatalytic methanol condensation cycle. Proc Natl Acad Sci 2014; 111(45): 15928-33.
[http://dx.doi.org/10.1073/pnas.1413470111] [PMID: 25355907]

[58]   Gu Y, Jiang Y, Yang S, Jiang W. Utilization of economical substrate-derived carbohydrates by solventogenic clostridia: Pathway dissection, regulation and engineering. Curr Opin Biotechnol 2014; 29: 124-31.
[http://dx.doi.org/10.1016/j.copbio.2014.04.004] [PMID: 24769507]

[59]   Broadbent JR, Neeno-Eckwall EC, Stahl B, *et al.* Analysis of the *Lactobacillus casei* supragenome and its influence in species evolution and lifestyle adaptation. BMC Genomics 2012; 13(1): 533.
[http://dx.doi.org/10.1186/1471-2164-13-533] [PMID: 23035691]

[60]   Hullahalli K, Rodrigues M, Schmidt BD, Li X, Bhardwaj P, Palmer KL. Comparative analysis of the orphan CRISPR2 locus in 242 *Enterococcus faecalis* strains. PLoS One 2015; 10(9): e0138890.
[http://dx.doi.org/10.1371/journal.pone.0138890] [PMID: 26398194]

[61]   Briner AE, Barrangou R. *Lactobacillus buchneri* genotyping on the basis of clustered regularly interspaced short palindromic repeat (CRISPR) locus diversity. Appl Environ Microbiol 2014; 80(3): 994-1001.
[http://dx.doi.org/10.1128/AEM.03015-13] [PMID: 24271175]

[62]   Tan Y, Shen J, Si T, Ho CL, Li Y, Dai L. Engineered live biotherapeutics: Progress and challenges. Biotechnol J 2020; 15(10): 2000155.
[http://dx.doi.org/10.1002/biot.202000155] [PMID: 32770635]

[63]   Paton AW, Jennings MP, Morona R, *et al.* Recombinant probiotics for treatment and prevention of enterotoxigenic *Escherichia coli* diarrhea. Gastroenterology 2005; 128(5): 1219-28.
[http://dx.doi.org/10.1053/j.gastro.2005.01.050] [PMID: 15887106]

[64]   Van Pijkeren JP, Barrangou R. Genome editing of food-grade *Lactobacilli* to develop therapeutic probiotics. Microbiol Spectr 2017; 5(5) 5.5.09.
[http://dx.doi.org/10.1128/microbiolspec.BAD-0013-2016] [PMID: 28959937]

[65]   Klotz C, Barrangou R. Engineering components of the *Lactobacillus* S-layer for biotherapeutic applications. Front Microbiol 2018; 9: 2264.
[http://dx.doi.org/10.3389/fmicb.2018.02264] [PMID: 30333802]

[66]   Barrangou R, Notebaart RA. CRISPR-directed microbiome manipulation across the food supply chain. Trends Microbiol 2019; 27(6): 489-96.
[http://dx.doi.org/10.1016/j.tim.2019.03.006] [PMID: 31003873]

[67]   Vercoe RB, Chang JT, Dy RL, *et al.* Cytotoxic chromosomal targeting by CRISPR/Cas systems can reshape bacterial genomes and expel or remodel pathogenicity islands. PLoS Genet 2013; 9(4): e1003454.
[http://dx.doi.org/10.1371/journal.pgen.1003454] [PMID: 23637624]

[68] Bikard D, Euler C, Jiang W, *et al.* Development of sequence-specific antimicrobials based on programmable CRISPR-Cas nucleases. Nat Biotechnol 2014; 32(11): 1146.
[http://dx.doi.org/10.1038/nbt.3043] [PMID: 25282355]

[69] Wu M, Li R. A novel role of the Type I CRISPR-Cas system in impairing host immunity by targeting endogenous genes. Am Assoc Immnol. 2016.
[http://dx.doi.org/10.4049/jimmunol.196.Supp.200.14]

[70] Hu M, Cherkaoui I, Misra S, Rutter GA. Functional genomics in pancreatic β cells: Recent advances in gene deletion and genome editing technologies for diabetes research. Front Endocrinol 2020; 11: 576632.
[http://dx.doi.org/10.3389/fendo.2020.576632] [PMID: 33162936]

[71] Wang CH, Lundh M, Fu A, *et al.* CRISPR-engineered human brown-like adipocytes prevent diet-induced obesity and ameliorate metabolic syndrome in mice. Sci Transl Med 2020; 12(558): eaaz8664.
[http://dx.doi.org/10.1126/scitranslmed.aaz8664] [PMID: 32848096]

[72] Kumar P, Malik YS, Ganesh B, *et al.* CRISPR-Cas system: An approach with potentials for COVID-19 diagnosis and therapeutics. Front Cell Infect Microbiol 2020; 10: 576875.
[http://dx.doi.org/10.3389/fcimb.2020.576875] [PMID: 33251158]

[73] Freije CA, Myhrvold C, Boehm CK, *et al.* Programmable inhibition and detection of RNA viruses using Cas13. Mol Cell 2019; 76(5): 826-837.e11.
[http://dx.doi.org/10.1016/j.molcel.2019.09.013] [PMID: 31607545]

[74] Wang G, Zhao N, Berkhout B, Das AT. CRISPR-Cas based antiviral strategies against HIV-1. Virus Res 2018; 244: 321-32.
[http://dx.doi.org/10.1016/j.virusres.2017.07.020] [PMID: 28760348]

[75] Arnoult N, Correia A, Ma J, *et al.* Regulation of DNA repair pathway choice in S and G2 phases by the NHEJ inhibitor CYREN. Nature 2017; 549(7673): 548-52.
[http://dx.doi.org/10.1038/nature24023] [PMID: 28959974]

[76] Selle K, Barrangou R. Harnessing CRISPR–Cas systems for bacterial genome editing. Trends Microbiol 2015; 23(4): 225-32.
[http://dx.doi.org/10.1016/j.tim.2015.01.008] [PMID: 25698413]

[77] Brissett NC, Doherty AJ. Repairing DNA double-strand breaks by the prokaryotic non-homologous end-joining pathway. Biochem Soc Trans 2009; 37(3): 539-45.
[http://dx.doi.org/10.1042/BST0370539] [PMID: 19442248]

[78] Gomaa AA, Klumpe HE, Luo ML, Selle K, Barrangou R, Beisel CL. Programmable removal of bacterial strains by use of genome-targeting CRISPR-Cas systems. MBio 2014; 5(1): e00928-13.
[http://dx.doi.org/10.1128/mBio.00928-13] [PMID: 24473129]

[79] Moeller R, Stackebrandt E, Reitz G, *et al.* Role of DNA repair by nonhomologous-end joining in *Bacillus subtilis* spore resistance to extreme dryness, mono- and polychromatic UV, and ionizing radiation. J Bacteriol 2007; 189(8): 3306-11.
[http://dx.doi.org/10.1128/JB.00018-07] [PMID: 17293412]

[80] Tao Y, Li X, Liu Y, *et al.* Structural analysis of Shu proteins reveals a DNA binding role essential for resisting damage. J Biol Chem 2012; 287(24): 20231-9.
[http://dx.doi.org/10.1074/jbc.M111.334698] [PMID: 22465956]

[81] Fu Y, Sander JD, Reyon D, Cascio VM, Joung JK. Improving CRISPR-Cas nuclease specificity using truncated guide RNAs. Nat Biotechnol 2014; 32(3): 279-84.
[http://dx.doi.org/10.1038/nbt.2808] [PMID: 24463574]

[82] Bae S, Park J, Kim JS. Cas-OFFinder: A fast and versatile algorithm that searches for potential off-target sites of Cas9 RNA-guided endonucleases. Bioinformatics 2014; 30(10): 1473-5.
[http://dx.doi.org/10.1093/bioinformatics/btu048] [PMID: 24463181]

[83]　Concordet JP, Haeussler M. CRISPOR: Intuitive guide selection for CRISPR/Cas9 genome editing experiments and screens. Nucleic Acids Res 2018; 46(W1): W242-5.
[http://dx.doi.org/10.1093/nar/gky354] [PMID: 29762716]

[84]　Chuai G, Ma H, Yan J, *et al.* DeepCRISPR: Optimized CRISPR guide RNA design by deep learning. Genome Biol 2018; 19(1): 80.
[http://dx.doi.org/10.1186/s13059-018-1459-4] [PMID: 29945655]

[85]　Ran FA, Hsu PD, Lin CY, *et al.* Double nicking by RNA-guided CRISPR-Cas9 for enhanced genome editing specificity. Cell 2013; 154(6): 1380-9.
[http://dx.doi.org/10.1016/j.cell.2013.08.021] [PMID: 23992846]

[86]　Lone BA, Karna SKL, Ahmad F, Shahi N, Pokharel YR. CRISPR/Cas9 system: A 859 bacterial tailor for genomic engineering. Genet Res Int 2018; 2018.

[87]　Yang Y, Xu J, Ge S, Lai L. CRISPR/Cas: Advances, limitations, and applications for precision cancer research. Front Med 2021; 8: 649896.
[http://dx.doi.org/10.3389/fmed.2021.649896] [PMID: 33748164]

[88]　Lee B, Lee K, Panda S, *et al.* Nanoparticle delivery of CRISPR into the brain rescues a mouse model of fragile X syndrome from exaggerated repetitive behaviours. Nat Biomed Eng 2018; 2(7): 497-507.
[http://dx.doi.org/10.1038/s41551-018-0252-8] [PMID: 30948824]

[89]　Zhang XH, Tee LY, Wang XG, Huang QS, Yang SH. Off-target effects in CRISPR/Cas9-mediated genome engineering. Mol Ther Nucleic Acids 2015; 4(11): e264.
[http://dx.doi.org/10.1038/mtna.2015.37] [PMID: 26575098]

[90]　Mougiakos I, Mohanraju P, Bosma EF, *et al.* Characterizing a thermostable Cas9 for bacterial genome editing and silencing. Nat Commun 2017; 8(1): 1647.
[http://dx.doi.org/10.1038/s41467-017-01591-4] [PMID: 29162801]

[91]　Yan WX, Hunnewell P, Alfonse LE, *et al.* Functionally diverse type V CRISPR-Cas systems. Science 2019; 363(6422): 88-91.
[http://dx.doi.org/10.1126/science.aav7271] [PMID: 30523077]

[92]　Shabbir MAB, Hao H, Shabbir MZ, *et al.* Survival and evolution of CRISPR–Cas system in prokaryotes and its applications. Front Immunol 2016; 7: 375.
[http://dx.doi.org/10.3389/fimmu.2016.00375] [PMID: 27725818]

[93]　Liang L, Liu R, Garst AD, *et al.* CRISPR enabled trackable genome engineering for isopropanol production in *Escherichia coli.* Metab Eng 2017; 41: 1-10.
[http://dx.doi.org/10.1016/j.ymben.2017.02.009] [PMID: 28216108]

[94]　Zhao M, Huang D, Zhang X, Koffas MAG, Zhou J, Deng Y. Metabolic engineering of *Escherichia coli* for producing adipic acid through the reverse adipate-degradation pathway. Metab Eng 2018; 47: 254-62.
[http://dx.doi.org/10.1016/j.ymben.2018.04.002] [PMID: 29625225]

[95]　Gao C, Wang S, Hu G, *et al.* Engineering *Escherichia coli* for malate production by integrating modular pathway characterization with CRISPRi-guided multiplexed metabolic tuning. Biotechnol Bioeng 2018; 115(3): 661-72.
[http://dx.doi.org/10.1002/bit.26486] [PMID: 29105733]

[96]　Kim SK, Han GH, Seong W, *et al.* CRISPR interference-guided balancing of a biosynthetic mevalonate pathway increases terpenoid production. Metab Eng 2016; 38: 228-40.
[http://dx.doi.org/10.1016/j.ymben.2016.08.006] [PMID: 27569599]

[97]　Wu J, Du G, Chen J, Zhou J. Enhancing flavonoid production by systematically tuning the central metabolic pathways based on a CRISPR interference system in *Escherichia coli.* Sci Rep 2015; 5(1): 13477.
[http://dx.doi.org/10.1038/srep13477] [PMID: 26323217]

[98]   Wen Z, Minton NP, Zhang Y, *et al.* Enhanced solvent production by metabolic engineering of a twin-clostridial consortium. Metab Eng 2017; 39: 38-48.
[http://dx.doi.org/10.1016/j.ymben.2016.10.013] [PMID: 27794465]

[99]   Wu Y, Chen T, Liu Y, *et al.* CRISPRi allows optimal temporal control of N-acetylglucosamine bioproduction by a dynamic coordination of glucose and xylose metabolism in *Bacillus subtilis.* Metab Eng 2018; 49: 232-41.
[http://dx.doi.org/10.1016/j.ymben.2018.08.012] [PMID: 30176395]

CHAPTER 3

Modulating the Gut Microbiome through Genome Editing for Alleviating Gut Dysbiosis

Atul R. Chavan[1,2,#]**, Maitreyee Pathak**[2,#]**, Hemant J. Purohit**[2] **and Anshuman A. Khardenavis**[1,2,*]

[1] *Academy of Scientific and Innovative Research (AcSIR), Ghaziabad-201002, India*

[2] *Environmental Biotechnology and Genomics Division (EBGD), CSIR–National Environmental Engineering Research Institute (NEERI), Nehru Marg, Nagpur-440020, Maharashtra, India*

Abstract: One of the components of the emerging lifestyle shows an exponential rise in the consumption of packaged or high-calorie food. This has caused an increase in the incidences of diseases which are considered to be a consequence of the changing lifestyle. It has been observed that these clinical conditions are linked with gut dysbiosis, and hence it has been proposed that by modulation of the composition of gut microbiota, the risk of such diseases can be lowered. Prebiotics and probiotics, in combination, possess tremendous potential for maintaining the homeostasis in individuals. In this chapter, a comparative assessment of CRISPR-mediated genome editing technique has been discussed with conventional omics tools and modelling approaches. These techniques substantially simplify the modification of target genome in complex microbial communities and could enhance their prebiotic and probiotic potential. The synthetic biology approach to microbiome therapies such as additive, subtractive, and modulatory therapies for curing gut dysbiosis are also discussed. The chapter is aimed at developing a better understanding about the role of CRISPR/Cas as a reliable technology that may be employed as a diagnostic tool for infectious disease diagnosis as well as its treatment. Although, the tool has already demonstrated its use in a wide range of genome editing and genetic engineering applications, additional study into its use in human genome editing and diagnostics is needed considering any potential side effects or ambiguities.

Keywords: Gut microbiome, Gut dysbiosis, Genome editing, Prebiotics and probiotics, CRISPR.

---

[*] **Corresponding author Anshuman A. Khardenavis:** Academy of Scientific and Innovative Research (AcSIR), Ghaziabad-201002, India & Environmental Biotechnology and Genomics Division (EBGD), CSIR–National Environmental Engineering Research Institute (NEERI), Nehru Marg, Nagpur-440020, Maharashtra, India; E-mail: aa_khardenavis@neeri.res.in
[#] Equal contribution

Prakash M. Halami & Aravind Sundararaman (Eds.)

INTRODUCTION

The concept of genome editing in molecular biology refers to the incorporation of changes in specific DNA sequences by insertion, deletion, or modification of the genome [1, 2]. Genes have been manipulated using this technique in a variety of ways, including modifying their nucleotide sequences and changing their expression. Several enzymes that have been employed for achieving the aforesaid activity include zinc finger nucleases [ZFN], transcription activator-like effector nucleases [TALEN], and homing meganucleases, all of which have demonstrated their efficiency, though they have to be reengineered for each target sequence [3]. However, the drawbacks of homologous recombination (HR) based genome editing, such as the large sample volume requirement and lower editing effectiveness, have restricted their wide-scale application [4]. These drawbacks of the conventional genome editing tools have been overcome by CRISPR (clustered regularly interspaced short palindromic repeats)-mediated genome editing techniques that have substantially simplified the targeted genome modification in complex organisms [1, 5]. Because of its ease of use, cost-effectiveness, and high efficacy of desired targeted changes, the CRISPR/Cas9 system has become the editing tool of choice in recent years in transformed cell cultures and secondary clones [6]. However, considering the inefficiencies and inaccuracies, the efficiencies of genome editing experiments need to be validated by Fragment analysis and/or Sanger sequencing by capillary electrophoresis [CE] [7, 8].

Among the different species of organisms, CRISPR/Cas9 has been utilised in plants, fungi, and mammals for genome editing [9], which has helped us to gain a better understanding of how a gene product contributes to an organism's development and disease [10]. In this chapter, the significance of the CRISPR/Cas system in microbiome editing has been discussed with reference to controlling the gene expression and regulation of metabolites and protein production in the gut. Metabolites such as prebiotics have the capacity to modulate the gut microbial community, which plays a critical role in various illness situations, thereby aiding in the maintenance of homeostasis of an individual. The chapter also highlights the application of CRISPR/Cas in precise diagnostics of diseases arising due to gut dysbiosis, and the detection of microorganisms and their underlying mechanisms suggested to be responsible for the disease conditions. Various platforms based on viruses have been developed for applying the CRISPR-Cas genome editing in gut microbiome studies, such as DNA endonuclease-targeted CRISPR trans reporter [DETECTR] HUDSON, SHERLOCK [Specific high-sensitivity enzymatic reporter unlocking] [11].

CRISPR AND GUT MICROBIOME MANAGEMENT BY PREBIOTICS, PROBIOTICS, AND SYNBIOTICS

The gut microbiome has an impact on the health status of an individual and has been associated with numerous diseases that are manifested in the form of inflammation and immunosenescence, leading to a condition termed gut dysbiosis. These effects are caused due to the immunomodulatory properties of gut microbiota, whose low diversity in aged people is characterized by facultative anaerobes and pathobionts that are responsible for the increase in the inflammatory signals [12 - 15]. During old age, a lower abundance of probiotic beneficial bacteria - *Bacteroides*, *Bifidobacteria*, and *Lactobacilli,* and an increase in abundance of *Enterococci, Coliforms,* and especially *Clostridium perfringens* and *C. difficile* were found to result in a reduction in the relative stability of the gut microbiome [16 - 18]. Similarly, other researchers observed a decrease in *Prevotella, Candida albicans, Streptococcus, Staphylococcus*, and *Faecalibacterium prausnitzii* and an increase in levels of *Ruminococcus* and *Atopobium* that established the relationship between frailty score and microbiota diversity in older people [19, 20]. Considering the above aspects of gut dysbiosis, methods for managing the gut microbiome are of paramount importance for maintaining the homeostasis of an individual. Prebiotics, probiotics, and synbiotics are concepts that have been researched extensively for their potential role in the management of the gut microbiome.

The term "Prebiotics" was coined by Gibson and Roberfroid [21] to describe a nutritional product and/or ingredient that selectively nourishes the gut microbiome, thereby providing health benefits to the host. Sources of natural prebiotics include garlic, onion, chicory root, barley, banana, tomato, and wheat, along with breast milk oligosaccharides that are the third-largest human milk component [22, 23]. Some of the synthetic oligosaccharides include fructooligosaccharides, galactooligosaccharides, xylooligosaccharides, maltooligosaccharides, and inulin [24, 25]. Prebiotics are food for probiotics. The term "Probiotics" which has its origin in Latin, "pro" meaning "for" and "bios" meaning "life", was coined by noble laureate Elie Winnerat at the beginning of the 20th century. Probiotics consist of live microorganisms that can give health benefits to the host when present in an adequate amount [26, 27]. They are non-pathogenic, non-toxic, and non-allergic [28] and are capable of inducing immunity against various diseases in addition to their ability to provide antimicrobial activity in the gut. Probiotic bacteria are capable of surviving and metabolizing carbohydrates, short-chain fatty acids, and bile acids in the upper gastrointestinal (GI) tract and are characterized by resistance to low pH, bile juice, and gastric acid [29]. Such functional food ingredients and supplements with health-enhancing effects have given rise to the concept of synbiotics that is defined as the synergistic effect of prebiotics and probiotics in enhancing the activity of beneficial microbiota in the gut [30, 31]. A synbiotic product aids the host in enhancing the viability and implantation of live microbial dietary supplements in the digestive tract. This beneficial effect is a result of a selective boost in the development and/or activation of the metabolic processes of one or a restricted number of health-promoting bacteria [32]. Thus, under the concept of synbiotics, the prebiotics and probiotics are made to function together with the co-administered microbe using the substrate preferentially [33].

Conventional Omics Tools for Improved Prebiotics Production

Agricultural waste represents a non-conventional and renewable source for a myriad of industrial applications. Different enzymes, such as cellulase, xylanase, and ligninase are essential for hydrolysis of the lignocellulosic material [34]. But the natural production of these hydrolytic enzymes in bacteria and fungi was restricted by their inducible nature. CRISPR/Cas technology has been demonstrated to achieve an improvement in enzyme induction in thermophilic bacteria such as *Myceliopthora thermophila* and *M. heterothallica*. Through non-homologous end-joining (NHEJ)-mediated processes, the CRISPR/Cas9 system was used for effectively altering the imported *amdS* gene in the genome [35]. The authors also inserted different genes of the cellulase pathway, such as *cre-1*, *res-1*, *gh1-1,* and *alp-1*, using CRISPR/Cas technology with a resultant multi-fold enhancement in cellulase production than the conventional methods. Similarly, CRISPR/Cas9 based genome editing was used for modifying the fungus *Talaromyces pinophilus* EMU for hyper production of extracellular cellulases [36]. β-galactosidase enzyme encoded by the *ganA* gene, is another important enzyme responsible for the degradation of the galactomannan fraction in agricultural residues for the production of prebiotic galactooligosaccharides (GOS). CRISPR vectors were used for higher production of β-galactosidase enzyme by integrating the *ganA* gene with glucitol promoter to make an expression cassette of a *Pgut-ganA* [37]. The aforesaid improved enzymes could be used for the enhanced saccharification of agricultural residues generating mono- or oligosaccharides that, in addition to their application in biofuel production, are also important raw materials for the production of prebiotic formulations.

Conventional Omics Tools for Improved Probiotics Production

The effectiveness and efficiency of probiotic bacteria can be upgraded and augmented by omics studies based on metagenomics, genomics, transcriptomics, nutritranscriptomics, nutrimetabolomics, *etc* [38 - 40]. Metagenomic sequencing has allowed the analysis of genetic structures for the identification of genetic changes [41]. Genomics and transcriptomics have found applications in various research and medical fields ranging from nutrigenomics, food, pharmaceutical industry, diagnostics, therapeutics, gene therapy applications, health and disease prevention, and developmental biology. Nutrigenomics comprises the three powerful omics analysis platforms in food and feed science [42, 43] that together can enable the researcher to reveal how genetic variations affect the metabolism by altering the genetic expression in animals [44]. The first goal of such studies is to analyse the characteristics of each feed component for enabling the effective utilization of food and food components as functional food factors in preventing lifestyle-related diseases [45]. In the nutritranscriptomic analysis, experimental animals or cultured cells are administered a nutrient or specific food component with altered composition by way of gavage, drinking water, or injection, followed by DNA microarray analysis for comparing the gene expression profiles between the control and experimental groups [46]. Similar applications are also reported in humans, although human studies are significantly hindered by ethical and technical issues. The application of genomics and transcriptomics to various aspects of food and feed has revolutionized the field of prebiotics and probiotics by providing deep insight into the biological processes involved in the effectiveness and potency of probiotic bacteria [47, 48].

CRISPR Based Genetic Manipulation in Prebiotics and Probiotics Production

The first processed food consumed by humans includes the various fermented foods and beverages such as cultured milk, yogurt, wine, beer, cider, tempeh, miso, kimchi, sauerkraut, and sausage that have been well-accepted due to their long shelf life, safety, and nutritional value [49]. The multipronged roles of starter cultures and probiotics in food processing include (a) food preservation by producing organic acids, hydrogen peroxide, and bacteriocins, (b) inhibiting pathogens for improved food safety, (c) enhancing the nutritional value and organoleptic qualities of food products, (d) delivery of live organisms to the gut for providing health benefits to the host [50, 51].

CRISPR/Cas9 is a novel, emerging gene-editing technique that can serve as an efficient alternative to the zinc-finger nucleases (ZFNs) and the transcription activator-like nucleases (TALENs) from the bacterium *Xanthomonas* for inducing targeted genetic modifications [52, 53]. The CRISPR/Cas technology has demonstrated the ability to carry out gene editing in target bacteria by various modes, *viz.*, transformation, conjugation, and transduction [54, 55]. It can improve the functionality of beneficial bacteria by increasing their antagonistic effect against the food spoilage bacteria, which highlights the potential application of CRISPR/Cas technology in the reduction of food spoilage [56].

The bacterial plasticity and its ability to reprogram the CRISPR/Cas systems have enabled genome enhancement to improve the probiotic traits in starter cultures. Lactic acid bacteria have a broad range of CRISPR/Cas systems which opens new avenues for utilizing these endogenous systems for genotyping and genome engineering, owing to their tendency to introduce mutations, insertions, or deletions [57]. From Table **1**, it is clear that such studies make use of CRISPR/Cas in gene editing by the introduction of site-specific mutations, gene deletions, and gene insertions which are then used to screen the desired mutants displaying enhanced activity of various functional genes in probiotic organisms [58, 59]. Similarly, the bacterial genome can be edited by using single-stranded DNA (ssDNA) recombination, double crossover recombination, or using Di-CRISPR (delta integration CRISPR/Cas) platform [58, 60 - 62]. Oh and van Pijkeren [62] reported a method for genome editing of *Lactobacillus reuteri* ATCC PTA 6475 possessing interesting immunomodulatory and antimicrobial properties. Genome editing by CRISPR/Cas9 has been demonstrated in *L. plantarum* wherein improvement in the recombination efficiency was achieved by using a recombination template on a plasmid rather than providing it as single-stranded DNA [63]. *Bifidobacterium* and *Lactobacillus* spp. are commercially used probiotic bacteria exerting a positive effect on human health. The modification of such probiotic bacteria for delivering vaccines or modulating the host's adaptive immune system could be made possible by CRISPR/Cas systems [64]. Clinical trials in rats and mice have shown promising effects of CRISPR based engineered probiotics on blood sugar reduction, reduction in the obesity related markers, as well as a reduction in the dietary oxalate [30, 65, 66]. Hence CRISPR based engineering has proven to be a very robust technique in engineering probiotic organisms which can be used in the detection as well as treatment of gut dysbiosis.

Table 1. CRISPR based genetic manipulation of probiotic bacteria.

Sr. No.	Genetically Engineered Probiotic Organism	Strategy	Effect	References
1.	*Lactococcus lactis*	Engineering of blood glucose reducing metabolite 'tagatose' in probiotic organisms	Oral administration of such engineered probiotics causes a decrease in blood sugar levels by converting galactose to tagatose	[65]
2.	*Lactobacillus gasseri*	Secretion of glucagon-like peptides	Conversion of the intestinal epithelial cells into glucose-responsive insulin-producing cells for treating diabetes	[73]

(Table 1) cont.....

Sr. No.	Genetically Engineered Probiotic Organism	Strategy	Effect	References
3.	*Bifidobacterium pseudocatenulatum CECT 7765*	Reduction in nitric oxide release by restoring the vascular function induced due to obesity	Decrease in obesity markers when fed with high-fat diet	[30]
4.	*Lactobacillus reuteri*	Genetically engineered *L. reuteri* used in the treatment of lactose intolerant patients as well as patients with phenylketonuria	Digestion of lactose using genetically engineered lactase enzyme in *L. reuteri* was improved and also genetically engineered phenylalanine ammonia lyase was successful in maintaining serum phenylalanine levels	[74]
5.	*Lactobacillus plantarum*	Enhanced secretion of oxalate decarboxylase in the intestine	Reduction in dietary oxalate was achieved, which inhibited the kidney stone formation	[66]
6.	*Lactobacillus reuteri* ATCC PTA 6475	Recombineering of CRISPR–Cas9 with single-stranded DNA (ssDNA) for identifying edited cells at high efficiencies	Potential change in the genome editing landscape of LAB, and other Gram-positive bacteria	[62]
7.	*Bifidobacterium animalis subsp. lactis DSM10140*	Double crossover recombination in *Bifidobacteria*	Surface structure characterization of the next generation probiotics for functional aspects considering their significance in interaction with the host	[60]
8.	*Saccharomyces cerevisiae*	Di-CRISPR (delta integration CRISPR/Cas) platform based on the CRISPR and CRISPR-associated systems (Cas) with specifically designed guide RNA sequences to target multiple delta sites in the yeast genome	Single-step, highly efficient and marker less, 18-copy genomic integration of a 24 kb combined xylose utilization and (R,R)-2,--butanediol (BDO) production pathway	[61]
9.	*Lactobacillus crispatus*	Generation of diverse mutations encompassing a 643-base pair (bp) deletion (100% efficiency), a stop codon insertion (36%), and a single nucleotide substitution (19%) in the exopolysaccharide priming-glycosyltransferase (*p-gtf*)	Harnessing of endogenous type I-E CRISPR/Cas system of *L. crispatus* for flexible and efficient genetic engineering encompassing insertions, deletions, and single base substitutions	[58]

(Table 1) cont.....

Sr. No.	Genetically Engineered Probiotic Organism	Strategy	Effect	References
10.	*Bacillus subtilis*	Knockdowns of every essential gene in *Bacillus subtilis* created using CRISPR interference and their phenotypes were probed	Provided the framework for systematic investigation of essential gene functions, which is applicable to a variety of microorganisms and used for comparative analysis	[75]
11.	*Clostridium beijerinckii*	Exploration of genome editing in microbes lacking sufficient genetic tools by applying the *Streptococcus pyogenes* CRISPR/Cas9 for genome editing in probiotic *Clostridium beijerinckii*	Efficient selection of desirable mutants through successful application of CRISPR/Cas9	[59]
12.	*Lactobacillus gasseri*	17 *L. gasseri* strains studied for the occurrence and activity of type 2 CRISPR/Cas system	Distribution and function of native type 2 CRISPR/Cas system in commensal species *L. gasseri* was reported, which contributes to the fundamental understanding of the CRISPR/Cas system in bacteria	[76]
13.	*Streptococcus thermophilus*	CRISPR-based genomic island screening to study excision events in probiotic bacteria	Deletions in genomic islands were identified, and the species were readily isolated using CRISPR/Cas screening applications which define minimal bacterial genomes, determine essential genes, and also characterize genetically heterogeneous bacterial populations	[77]
14.	*Lactobacillus plantarum*	Comparison of two methods in different strains of *L. plantarum*: one utilizing oligonucleotide donor as well as an inducible DNA recombinase and one utilizing a plasmid-encoded recombineering template	Both methods showed success in editing the same site across multiple strains of *L. plantarum*	[78]

(Table 1) cont.....

Sr. No.	Genetically Engineered Probiotic Organism	Strategy	Effect	References
15.	*Lactobacillus casei*	A rapid and precise genome editing plasmid, pLCNICK, established for *L. casei* genome engineering based on CRISPR/Cas9D10A	Efficient single-gene deletion and insertion accomplished by one-step transformation with a reduction in cycle time to 9 days. Limitations of previous methods were overcome by rapid and precise chromosomal manipulation in *L. casei,* thus enabling a comprehensive investigation of *L. casei*	[79]

MICROBIOME EDITING USING CRISPR

Synthetic Biology Approaches to Microbiome Therapies

Synthetic biology approaches aim to design a cell in such a way that it achieves programmed cell behaviour using either synthetic or natural biological components. Synthetic biological approaches such as genome editing, quorum sensing, and CRISPR/Cas enable researchers to study the structure-function relationships between microbiome and further engineer novel biotic therapeutics. Such synthetic components expand the potential of engineered microbe to respond to its local environment [67, 68].

Synthetic biology involves 3 types of approaches for editing the microbiome (Fig. **1**); i) Additive therapies in which host microbiota is augmented with consortia or individual strains of bacterial species, ii) Subtractive therapies involving the elimination of unwanted pathogens of microbiome, iii) Modulatory therapies used to alter the activity as well as composition of the microbiome which involves changes in metabolites as well as protein production of the microbiome [69, 70].

CRISPR Based Genetic Manipulation for Gut Microbiome Engineering

Initially, precise genome editing using CRISPR/Cas9 was done in *Streptococcus pyogenes*, where Cas9 protein from *S. pyogenes* was integrated into a pathogenic strain of *S. pneumoniae* present in the sinuses, respiratory tract, and nasal cavities with the aim of targeting the antibiotic resistance cassette. Most bacteria could not survive because of the CRISPR/Cas9 activity, but other bacteria which did not have an antibiotic resistance cassette survived through homologous recombination. This method was used in various other bacteria with success except in *E. coli,* where phage λ was used to enhance the Cas9 mediated break repair as well as editing [71]. Microbiome editing using CRISPR/Cas9 is dependent on the guide RNA expression, Lambda genes from several plasmids, and Cas9. Short ssDNA, long dsDNA or DNA cloned into a plasmid could be provided as templates which are interrupted by the insertion of Cas9 at the desired position leading to cell death unless repair or modification occurred in the target DNA. The technique of RNA-guided DNA cleavage for genome editing in cells was useful in creating mutant libraries in organisms possessing high recombinase activity as well as such organisms could be efficiently transformed. Probiotics can be engineered using such CRISPR mediated transformation to correct gut dysbiosis. Recombineering techniques using oligonucleotides have been recently upgraded using CRISPR/Cas9 in the probiotic lactic acid bacteria (LAB), giving rise to new opportunities in engineering LABs [62, 72]. Some clinical trials have shown promising results in the engineering of such probiotic bacteria for supplementation or replacement of an enzyme thus offering an approach for developing recombinant probiotics that would be helpful in correcting gut dysbiosis.

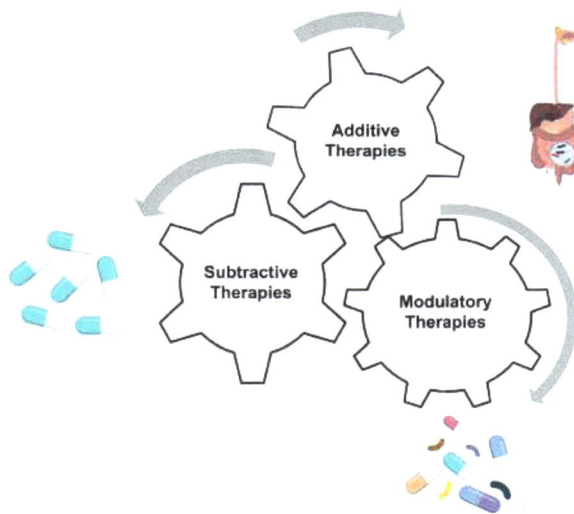

Fig. (1). Synthetic biology approaches for editing the microbiome.

CRISPR/Cas9 as a Portable Diagnostic Tool for Detecting Diseases

Gene editing using CRISPR/Cas proteins has transformed the field of molecular diagnostics in various infectious diseases by the detection of various viruses. Cas proteins generate CRISPR RNA (crRNA) and encourage adaptive immunity through it. During this process, the foreign genetic material is integrated into the

CRISPR array. Pre-crRNA is then transcribed and processed into a mature form, which escorts the Cas proteins for cleavage of complementary sequences of infectious elements for their degradation and elimination. Such a mechanism of CRISPR/Cas based genome editing has revolutionized the efforts to defeat various infectious agents [55, 80]. CRISPR/Cas12a proteins create trans-cleavage of dsDNA (double stranded DNA) into ssDNA [single stranded DNA]. This phenomenon was applied in the development of specific and robust tests for detection of carcinoma associated HPV as well as African swine fever virus, which is known as DNA endonuclease-targeted CRISPR trans reporter [DETECTR] [11]. The method with the acronym "HUDSON", which stands for 'heating unextracted diagnostic samples to obliterate nucleases' was developed to stop the degradation of nucleic acids from clinical specimens by applying heat and chemicals for inactivating the ribonucleases (RNases) in the body, further disrupting the viral envelope and releasing the nucleic acids [81]. Another tool for nucleic acid detection based on Cas9 similar to HUDSON was developed, which was known as SHERLOCK [Specific high-sensitivity enzymatic reporter unlocking]. In this method, RNA was coupled to a fluorescent reporter whose cleavage produced a fluorescent signal that was further amplified through enzymatic activity, thus increasing test sensitivity [82]. The current pandemic of COVID-19 has encouraged researchers to develop new diagnostic approaches which should overcome the disadvantages of RT-PCR. New protocols using diagnostic strips for detection of COVID-19 on paper readouts have been developed based on SHERLOCK which have reduced the time of detection while at the same time increasing the specificity [83].

CONCLUSION

Correction of gut dysbiosis for the treatment of gut-related metabolic disorders is now considered an auspicious approach encompassing a host of genetically engineered probiotic organisms, and their introduction into the host for fast and efficient effect. Among these technologies, CRISPR-Cas based genome editing is a robust technique that has the potential for use, both as a diagnostic tool as well as a therapy for detecting infectious diseases. A different and broad range of applications of CRISPR makes it a convenient and beneficial tool for microbiome editing wherein, synthetic biological approaches are used to design probiotics and edit the microbiome to correct gut dysbiosis. To conclude, in this chapter, the authors have explored and reviewed the application of CRISPR/Cas technology with special emphasis on the role of CRISPR/Cas technology in genome editing for the correction of gut dysbiosis by modulation of the gut microbiome through the use of prebiotics and probiotics.

ACKNOWLEDGEMENTS

All the authors have contributed equally to the manuscript.

The authors would like to acknowledge Director, CSIR-NEERI and AcSIR-NEERI for providing essential resources for the research work. The manuscript has been internally checked for plagiarism through iThenticate software and assigned the KRC No. CSIR-NEERI/KRC/2021/July/EBGD/3. ARC is thankful to CSIR for Senior Research Fellowship for carrying out this research.

REFERENCES

[1]   Zhang N, Roberts HM, Van Eck J, Martin GB. Generation and molecular characterization of CRISPR/Cas9-induced mutations in 63 immunity-associated genes in tomato reveals specificity and a range of gene modifications. Front Plant Sci 2020; 11: 10.
        [http://dx.doi.org/10.3389/fpls.2020.00010] [PMID: 32117361]

[2]   Han HA, Pang JKS, Soh BS. Mitigating off-target effects in CRISPR/Cas9-mediated *in vivo* gene editing. J Mol Med 2020; 98(5): 615-32.
        [http://dx.doi.org/10.1007/s00109-020-01893-z] [PMID: 32198625]

[3]   Watanabe T, Ochiai H, Sakuma T, *et al.* Non-transgenic genome modifications in a hemimetabolous insect using zinc-finger and TAL effector nucleases. Nat Commun 2012; 3(1): 1017.
        [http://dx.doi.org/10.1038/ncomms2020] [PMID: 22910363]

[4]   Ghosh D, Venkataramani P, Nandi S, Bhattacharjee S. CRISPR–Cas9 a boon or bane: The bumpy road ahead to cancer therapeutics. Cancer Cell Int 2019; 19(1): 12.
        [http://dx.doi.org/10.1186/s12935-019-0726-0] [PMID: 30636933]

[5]   Sander JD, Joung JK. CRISPR-Cas systems for editing, regulating and targeting genomes. Nat Biotechnol 2014; 32(4): 347-55.
        [http://dx.doi.org/10.1038/nbt.2842] [PMID: 24584096]

[6]   Liu H, Ding Y, Zhou Y, Jin W, Xie K, Chen LL. CRISPR-P 2.0: An improved CRISPR-Cas9 tool for genome editing in plants. Mol Plant 2017; 10(3): 530-2.
        [http://dx.doi.org/10.1016/j.molp.2017.01.003] [PMID: 28089950]

[7]   Grohmann L, Keilwagen J, Duensing N, *et al.* Detection and identification of genome editing in plants: Challenges and opportunities. Front Plant Sci 2019; 10: 236.
        [http://dx.doi.org/10.3389/fpls.2019.00236] [PMID: 30930911]

[8]   Kosicki M, Tomberg K, Bradley A. Repair of double-strand breaks induced by CRISPR–Cas9 leads to large deletions and complex rearrangements. Nat Biotechnol 2018; 36(8): 765-71.
        [http://dx.doi.org/10.1038/nbt.4192] [PMID: 30010673]

[9]   Kirchner M, Schneider S. CRISPR-Cas: From the bacterial adaptive immune system to a versatile tool for genome engineering. Angew Chem Int Ed 2015; 54(46): 13508-14.
        [http://dx.doi.org/10.1002/anie.201504741] [PMID: 26382836]

[10]   Hajiahmadi Z, Movahedi A, Wei H, *et al.* Strategies to increase on-target and reduce off-target effects of the CRISPR/Cas9 system in plants. Int J Mol Sci 2019; 20(15): 3719.
        [http://dx.doi.org/10.3390/ijms20153719] [PMID: 31366028]

[11]   Kim HR, Lee HM, Yu HC, *et al.* Biodegradation of polystyrene by Pseudomonas sp. isolated from the gut of superworms *(larvae of Zophobas atratus).* Environ Sci Technol 2020; 54(11): 6987-96.
        [http://dx.doi.org/10.1021/acs.est.0c01495] [PMID: 32374590]

[12]   Claesson MJ, Jeffery IB, Conde S, *et al.* Gut microbiota composition correlates with diet and health in the elderly. Nature 2012; 488(7410): 178-84.

[http://dx.doi.org/10.1038/nature11319] [PMID: 22797518]

[13]   Biagi E, Candela M, Fairweather-Tait S, Franceschi C, Brigidi P. Ageing of the human metaorganism: The microbial counterpart. Age 2012; 34(1): 247-67.
[http://dx.doi.org/10.1007/s11357-011-9217-5] [PMID: 21347607]

[14]   Biagi E, Nylund L, Candela M, *et al.* Through ageing, and beyond: gut microbiota and inflammatory status in seniors and centenarians. PLoS One 2010; 5(5): e10667.
[http://dx.doi.org/10.1371/journal.pone.0010667] [PMID: 20498852]

[15]   Hornef M. Pathogens, commensal symbionts, and pathobionts: discovery and functional effects on the host. ILAR J 2015; 56(2): 159-62.
[http://dx.doi.org/10.1093/ilar/ilv007] [PMID: 26323625]

[16]   Mitsuoka T. *Bifidobacteria* and their role in human health. J Ind Microbiol 1990; 6(4): 263-7.
[http://dx.doi.org/10.1007/BF01575871]

[17]   O'Toole PW, Jeffery IB. Gut microbiota and aging. Science 2015; 350(6265): 1214-5.
[http://dx.doi.org/10.1126/science.aac8469] [PMID: 26785481]

[18]   Rea MC, O'Sullivan O, Shanahan F, *et al. Clostridium difficile* carriage in elderly subjects and associated changes in the intestinal microbiota. J Clin Microbiol 2012; 50(3): 867-75.
[http://dx.doi.org/10.1128/JCM.05176-11] [PMID: 22162545]

[19]   Mariat D, Firmesse O, Levenez F, *et al.* The firmicutes/bacteroidetes ratio of the human microbiota changes with age. BMC Microbiol 2009; 9(1): 123.
[http://dx.doi.org/10.1186/1471-2180-9-123] [PMID: 19508720]

[20]   van Tongeren SP, Slaets JPJ, Harmsen HJM, Welling GW. Fecal microbiota composition and frailty. Appl Environ Microbiol 2005; 71(10): 6438-42.
[http://dx.doi.org/10.1128/AEM.71.10.6438-6442.2005] [PMID: 16204576]

[21]   Gibson GR, Hutkins R, Sanders ME, *et al.* Expert consensus document: The international scientific association for probiotics and prebiotics (ISAPP) consensus statement on the definition and scope of prebiotics. Nat Rev Gastroenterol Hepatol 2017; 14(8): 491-502.
[http://dx.doi.org/10.1038/nrgastro.2017.75] [PMID: 28611480]

[22]   Vinayak A, Mudgal G, Sharma S, Singh GB. Prebiotics for probiotics. Advances in Probiotics for Sustainable Food and Medicine 2021; pp. 63-82.

[23]   Mohanty D, Misra S, Mohapatra S, Sahu PS. Prebiotics and synbiotics: Recent concepts in nutrition. Food Biosci 2018; 26: 152-60.
[http://dx.doi.org/10.1016/j.fbio.2018.10.008]

[24]   Vera C, Illanes A, Guerrero C. Enzymatic production of prebiotic oligosaccharides. Curr Opin Food Sci 2021; 37: 160-70.
[http://dx.doi.org/10.1016/j.cofs.2020.10.013]

[25]   Markowiak P, Śliżewska K. The role of probiotics, prebiotics and synbiotics in animal nutrition. Gut Pathog 2018; 10(1): 21.
[http://dx.doi.org/10.1186/s13099-018-0250-0] [PMID: 29930711]

[26]   Pradhan D, Mallappa RH, Grover S. Comprehensive approaches for assessing the safety of probiotic bacteria. Food Control 2020; 108: 106872.
[http://dx.doi.org/10.1016/j.foodcont.2019.106872]

[27]   Gasbarrini G, Bonvicini F, Gramenzi A. Probiotics history. J Clin Gastroenterol 2016; 50 (2): S116-9.
[http://dx.doi.org/10.1097/MCG.0000000000000697] [PMID: 27741152]

[28]   Kerry RG, Patra JK, Gouda S, Park Y, Shin HS, Das G. Benefaction of probiotics for human health: A review. J Food Drug Anal 2018; 26(3): 927-39.
[PMID: 29976412]

[29]   Pandey KR, Naik SR, Vakil BV. Probiotics, prebiotics and synbiotics : A review. J Food Sci Technol

2015; 52(12): 7577-87.
[http://dx.doi.org/10.1007/s13197-015-1921-1] [PMID: 26604335]

[30]   Mauricio MD, Serna E, Fernández-Murga ML, *et al. Bifidobacterium pseudocatenulatum* CECT 7765 supplementation restores altered vascular function in an experimental model of obese mice. Int J Med Sci 2017; 14(5): 444-51.
[http://dx.doi.org/10.7150/ijms.18354] [PMID: 28539820]

[31]   Shoaib M, Shehzad A, Omar M, *et al.* Inulin: Properties, health benefits and food applications. Carbohydr Polym 2016; 147: 444-54.
[http://dx.doi.org/10.1016/j.carbpol.2016.04.020] [PMID: 27178951]

[32]   Kolida S, Gibson GR. Synbiotics in health and disease. Annu Rev Food Sci Technol 2011; 2(1): 373-93.
[http://dx.doi.org/10.1146/annurev-food-022510-133739] [PMID: 22129388]

[33]   Swanson KS, Gibson GR, Hutkins R, *et al.* The international scientific association for probiotics and prebiotics (ISAPP) consensus statement on the definition and scope of synbiotics. Nat Rev Gastroenterol Hepatol 2020; 17(11): 687-701.
[http://dx.doi.org/10.1038/s41575-020-0344-2] [PMID: 32826966]

[34]   Nakhate SP, Gupta RK, Poddar BJ, *et al.* Influence of lignin level of raw material on anaerobic digestion process in reorganization and performance of microbial community. Int J Environ Sci Technol 2022; 19(3): 1819-36.
[http://dx.doi.org/10.1007/s13762-021-03141-4]

[35]   Liu Q, Gao R, Li J, *et al.* Development of a genome-editing CRISPR/Cas9 system in thermophilic fungal *Myceliophthora* species and its application to hyper-cellulase production strain engineering. Biotechnol Biofuels 2017; 10(1): 1-4.
[http://dx.doi.org/10.1186/s13068-016-0693-9] [PMID: 28053662]

[36]   Manglekar RR, Geng A. CRISPR-Cas9-mediated seb1 disruption in *Talaromyces pinophilus* EMU for its enhanced cellulase production. Enzyme Microb Technol 2020; 140: 109646.
[http://dx.doi.org/10.1016/j.enzmictec.2020.109646] [PMID: 32912697]

[37]   Watzlawick H, Altenbuchner J. Multiple integration of the gene ganA into the *Bacillus subtilis* chromosome for enhanced β-galactosidase production using the CRISPR/Cas9 system. AMB Express 2019; 9(1): 158.
[http://dx.doi.org/10.1186/s13568-019-0884-4] [PMID: 31571017]

[38]   Zoumpopoulou G, Kazou M, Alexandraki V, *et al.* Probiotics and prebiotics: An overview on recent trends. Probiotics and Prebiotics in Animal Health and Food Safety 2018; pp. 1-34.

[39]   Purohit HJ, Kapley A, Khardenavis A, Qureshi A, Dafale NA. Insights in waste management bioprocesses using genomic tools. Adv Appl Microbiol 2016; 97: 121-70.
[http://dx.doi.org/10.1016/bs.aambs.2016.09.002] [PMID: 27926430]

[40]   Sánchez B, Delgado S, Blanco-Míguez A, Lourenço A, Gueimonde M, Margolles A. Probiotics, gut microbiota, and their influence on host health and disease. Mol Nutr Food Res 2017; 61(1): 1600240.
[http://dx.doi.org/10.1002/mnfr.201600240] [PMID: 27500859]

[41]   Yadav S, Kapley A. Exploration of activated sludge resistome using metagenomics. Sci Total Environ 2019; 692: 1155-64.
[http://dx.doi.org/10.1016/j.scitotenv.2019.07.267] [PMID: 31539947]

[42]   Gujar VV, Fuke P, Khardenavis AA, Purohit HJ. Draft genome sequence of *Penicillium chrysogenum* strain HKF2, a fungus with potential for production of prebiotic synthesizing enzymes. 3 Biotech 2018; 8: 1-5.

[43]   Cole MB, Augustin MA, Robertson MJ, Manners JM. The science of food security. NPJ Sci Food 2018; 2(1): 14.
[http://dx.doi.org/10.1038/s41538-018-0021-9] [PMID: 31304264]

[44] Benítez R, Núñez Y, Óvilo C, Ovilo C. Nutrigenomics in farm animals. J Invest Genomics 2017; 4: 1.

[45] Rana S, Kumar S, Rathore N, Padwad Y, Bhushan S. Nutrigenomics and its impact on life style associated metabolic diseases. Curr Genomics 2016; 17(3): 261-78.
[http://dx.doi.org/10.2174/1389202917666160202220422] [PMID: 27252592]

[46] Odriozola L, Corrales FJ. Discovery of nutritional biomarkers: Future directions based on omics technologies. Int J Food Sci Nutr 2015; 66(Suppl 1): S31-40.
[http://dx.doi.org/10.3109/09637486.2015.1038224]

[47] Börner RA, Kandasamy V, Axelsen AM, Nielsen AT, Bosma EF. Genome editing of lactic acid bacteria: Opportunities for food, feed, pharma and biotech. FEMS Microbiol Lett 2019; 366(1): fny291.
[http://dx.doi.org/10.1093/femsle/fny291] [PMID: 30561594]

[48] Hasin Y, Seldin M, Lusis A. Multi-omics approaches to disease. Genome Biol 2017; 18(1): 83.
[http://dx.doi.org/10.1186/s13059-017-1215-1] [PMID: 28476144]

[49] Tamang JP, Cotter PD, Endo A, *et al.* Fermented foods in a global age: East meets West. Compr Rev Food Sci Food Saf 2020; 19(1): 184-217.
[http://dx.doi.org/10.1111/1541-4337.12520] [PMID: 33319517]

[50] Bourdichon F, Casaregola S, Farrokh C, *et al.* Food fermentations: Microorganisms with technological beneficial use. Int J Food Microbiol 2012; 154(3): 87-97.
[http://dx.doi.org/10.1016/j.ijfoodmicro.2011.12.030] [PMID: 22257932]

[51] Özogul F, Hamed I. The importance of lactic acid bacteria for the prevention of bacterial growth and their biogenic amines formation: A review. Crit Rev Food Sci Nutr 2018; 58(10): 1660-70.
[http://dx.doi.org/10.1080/10408398.2016.1277972] [PMID: 28128651]

[52] Bortesi L, Fischer R. The CRISPR/Cas9 system for plant genome editing and beyond. Biotechnol Adv 2015; 33(1): 41-52.
[http://dx.doi.org/10.1016/j.biotechadv.2014.12.006] [PMID: 25536441]

[53] Gupta D, Bhattacharjee O, Mandal D, *et al.* CRISPR-Cas9 system: A new-fangled dawn in gene editing. Life Sci 2019; 232: 116636.
[http://dx.doi.org/10.1016/j.lfs.2019.116636] [PMID: 31295471]

[54] Glass Z, Lee M, Li Y, Xu Q. Engineering the delivery system for CRISPR-based genome editing. Trends Biotechnol 2018; 36(2): 173-85.
[http://dx.doi.org/10.1016/j.tibtech.2017.11.006] [PMID: 29305085]

[55] Hille F, Richter H, Wong SP, Bratovič M, Ressel S, Charpentier E. The biology of CRISPR-Cas: Backward and forward. Cell 2018; 172(6): 1239-59.
[http://dx.doi.org/10.1016/j.cell.2017.11.032] [PMID: 29522745]

[56] Stout E, Klaenhammer T, Barrangou R. CRISPR-Cas technologies and applications in food bacteria. Annu Rev Food Sci Technol 2017; 8(1): 413-37.
[http://dx.doi.org/10.1146/annurev-food-072816-024723] [PMID: 28245154]

[57] Barrangou R, Van Pijkeren JP. Exploiting CRISPR–Cas immune systems for genome editing in bacteria. Curr Opin Biotechnol 2016; 37: 61-8.
[http://dx.doi.org/10.1016/j.copbio.2015.10.003] [PMID: 26629846]

[58] Hidalgo-Cantabrana C, Goh YJ, Pan M, Sanozky-Dawes R, Barrangou R. Genome editing using the endogenous type I CRISPR-Cas system in *Lactobacillus crispatus*. Proc Natl Acad Sci 2019; 116(32): 15774-83.
[http://dx.doi.org/10.1073/pnas.1905421116] [PMID: 31341082]

[59] Wang Y, Zhang ZT, Seo SO, *et al.* Bacterial genome editing with CRISPR-Cas9: deletion, integration, single nucleotide modification, and desirable "clean" mutant selection in *Clostridium beijerinckii* as an example. ACS Synth Biol 2016; 5(7): 721-32.

[http://dx.doi.org/10.1021/acssynbio.6b00060] [PMID: 27115041]

[60] Castro-Bravo N, Wells JM, Margolles A, Ruas-Madiedo P. Interactions of surface exopolysaccharides from *Bifidobacterium* and Lactobacillus within the intestinal environment. Front Microbiol 2018; 9: 2426.
[http://dx.doi.org/10.3389/fmicb.2018.02426] [PMID: 30364185]

[61] Shi S, Liang Y, Zhang MM, Ang EL, Zhao H. A highly efficient single-step, markerless strategy for multi-copy chromosomal integration of large biochemical pathways in *Saccharomyces cerevisiae*. Metab Eng 2016; 33: 19-27.
[http://dx.doi.org/10.1016/j.ymben.2015.10.011] [PMID: 26546089]

[62] Oh JH, Van Pijkeren JP. CRISPR–Cas9-assisted recombineering in *Lactobacillus reuteri*. Nucleic Acids Res 2014; 42(17): e131.
[http://dx.doi.org/10.1093/nar/gku623] [PMID: 25074379]

[63] Leenay RT, Vento JM, Shah M, Martino ME, Leulier F, Beisel CL. Streamlined, recombinase-free genome editing with CRISPR-Cas9 in *Lactobacillus plantarum* reveals barriers to efficient editing. BioRxiv 2018; 352039.

[64] Hidalgo-Cantabrana C, O'Flaherty S, Barrangou R. CRISPR-based engineering of next-generation lactic acid bacteria. Curr Opin Microbiol 2017; 37: 79-87.
[http://dx.doi.org/10.1016/j.mib.2017.05.015] [PMID: 28622636]

[65] Rhimi M, Bermudez-Humaran LG, Huang Y, *et al.* The secreted L-arabinose isomerase displays anti-hyperglycemic effects in mice. Microb Cell Fact 2015; 14(1): 204.
[http://dx.doi.org/10.1186/s12934-015-0391-5] [PMID: 26691177]

[66] Sasikumar P, Gomathi S, Anbazhagan K, *et al.* Recombinant *Lactobacillus plantarum* expressing and secreting heterologous oxalate decarboxylase prevents renal calcium oxalate stone deposition in experimental rats. J Biomed Sci 2014; 21(1): 86.
[http://dx.doi.org/10.1186/s12929-014-0086-y] [PMID: 25175550]

[67] Park M, Tsai SL, Chen W. Microbial biosensors: Engineered microorganisms as the sensing machinery. Sensors 2013; 13(5): 5777-95.
[http://dx.doi.org/10.3390/s130505777] [PMID: 23648649]

[68] Rong G, Corrie SR, Clark HA. *In vivo* biosensing: Progress and perspectives. ACS Sens 2017; 2(3): 327-38.
[http://dx.doi.org/10.1021/acssensors.6b00834] [PMID: 28723197]

[69] Bober JR, Beisel CL, Nair NU. Synthetic biology approaches to engineer probiotics and members of the human microbiota for biomedical applications. Annu Rev Biomed Eng 2018; 20(1): 277-300.
[http://dx.doi.org/10.1146/annurev-bioeng-062117-121019] [PMID: 29528686]

[70] Ramachandran G, Bikard D. Editing the microbiome the CRISPR way. Philosophical Transactions of the Royal Society B. 1772; 374: p. (1772); 20180103.

[71] Jiang W, Bikard D, Cox D, Zhang F, Marraffini LA. RNA-guided editing of bacterial genomes using CRISPR-Cas systems. Nat Biotechnol 2013; 31(3): 233-9.
[http://dx.doi.org/10.1038/nbt.2508] [PMID: 23360965]

[72] Van Pijkeren JP, Britton RA. Precision genome engineering in lactic acid bacteria. Microb Cell Fact 2014; 13(Suppl 1) (1): S10.
[http://dx.doi.org/10.1186/1475-2859-13-S1-S10] [PMID: 25185700]

[73] Duan FF, Liu JH, March JC. Engineered commensal bacteria reprogram intestinal cells into glucose-responsive insulin-secreting cells for the treatment of diabetes. Diabetes 2015; 64(5): 1794-803.
[http://dx.doi.org/10.2337/db14-0635] [PMID: 25626737]

[74] Durrer KE, Allen MS, Hunt von Herbing I. Genetically engineered probiotic for the treatment of phenylketonuria (PKU); assessment of a novel treatment *in vitro* and in the PAHenu2 mouse model of PKU. PLoS One 2017; 12(5): e0176286.

[http://dx.doi.org/10.1371/journal.pone.0176286] [PMID: 28520731]

[75]   Peters JM, Colavin A, Shi H, *et al.* A comprehensive, CRISPR-based functional analysis of essential genes in bacteria. Cell 2016; 165(6): 1493-506.
[http://dx.doi.org/10.1016/j.cell.2016.05.003] [PMID: 27238023]

[76]   Sanozky-Dawes R, Selle K, O'Flaherty S, Klaenhammer T, Barrangou R. Occurrence and activity of a type II CRISPR-Cas system in *Lactobacillus gasseri*. Microbiology 2015; 161(9): 1752-61.
[http://dx.doi.org/10.1099/mic.0.000129] [PMID: 26297561]

[77]   Selle K, Klaenhammer TR, Barrangou R. CRISPR-based screening of genomic island excision events in bacteria. Proc Natl Acad Sci 2015; 112(26): 8076-81.
[http://dx.doi.org/10.1073/pnas.1508525112] [PMID: 26080436]

[78]   Leenay RT, Vento JM, Shah M, Martino ME, Leulier F, Beisel CL. Genome editing with CRISPR-Cas9 in *Lactobacillus plantarum* revealed that editing outcomes can vary across strains and between methods. Biotechnol J 2019; 14(3): 1700583.
[http://dx.doi.org/10.1002/biot.201700583] [PMID: 30156038]

[79]   Song X, Huang H, Xiong Z, Ai L, Yang S. CRISPR-Cas9D10A nickase-assisted genome editing in *Lactobacillus casei*. Appl Environ Microbiol 2017; 83(22): e01259-17.
[http://dx.doi.org/10.1128/AEM.01259-17] [PMID: 28864652]

[80]   Piepenburg O, Williams CH, Stemple DL, Armes NA. DNA detection using recombination proteins. PLoS Biol 2006; 4(7): e204.
[http://dx.doi.org/10.1371/journal.pbio.0040204] [PMID: 16756388]

[81]   Myhrvold C, Freije CA, Gootenberg JS, *et al.* Field-deployable viral diagnostics using CRISPR-Cas13. Science 2018; 360(6387): 444-8.
[http://dx.doi.org/10.1126/science.aas8836] [PMID: 29700266]

[82]   Cox DBT, Gootenberg JS, Abudayyeh OO, *et al.* RNA editing with CRISPR-Cas13. Science 2017; 358(6366): 1019-27.
[http://dx.doi.org/10.1126/science.aaq0180] [PMID: 29070703]

[83]   Zhang F, Abudayyeh OO, Gootenberg JS. A protocol for detection of COVID-19 using CRISPR diagnostics. Broad Institute 2020; p. 8.

CHAPTER 4

Bifidobacterial Genome Editing for Potential Probiotic Development

Kriti Ghatani[1,*], Shankar Prasad Sha[2], Subarna Thapa[1], Priya Chakraborty[1] and Sagnik Sarkar[1]

[1] *Department of Food Technology, University of North Bengal, Raja Rammohunpur, Darjeeling, West Bengal, 734013, India*

[2] *Department of Botany, Food Microbiology Lab, Kurseong College, University of North Bengal, Dow Hill Road, Kurseong, Darjeeling 7342003, West Bengal, India*

Abstract: Genome editing is a promising tool in the era of modern biotechnology that can alter the DNA of many organisms. It is now extensively used in various industries to obtain the well-desired and enhanced characteristics to improve the yield and nutritional quality of products. The positive health attributes of *Bifidobacteria*, such as prevention of diarrhoea, reduction of ulcerative colitis, prevention of necrotizing enterocolitis, *etc.*, have shown promising reports in many clinical trials. The potential use of *Bifidobacteria* as starter or adjunct cultures has become popular. Currently, *Bifidobacterium bifidum, B. adolescentis, B. breve, B. infantis, B. longum,* and *B. lactis* find a significant role in the development of probiotic fermented dairy products. However, *Bifidobacteria*, one of the first colonizers of the human GI tract and an indicator of the health status of an individual, has opened new avenues for research and, thereby, its application. Besides this, the GRAS/QPS (Generally Regarded as Safe/Qualified Presumption of Safety) status of *Bifidobacteria* makes it safe for use. They belong to the subgroup (which are the fermentative types that are primarily found in the natural cavities of humans and animals) of *Actinomycetes*. *B. lactis* has been used industrially in fermented foods, such as yogurt, cheese, beverages, sausages, infant formulas, and cereals. In the present book chapter, the authors tried to explore the origin, health attributes, and various genetic engineering tools for genome editing of *Bifidobacteria* for the development of starter culture for dairy and non-dairy industrial applications as well as probiotics.

Keywords: *Bifidobacteria*, CRISPR-Cas, Genome editing, IPSD (Inducible Plasmid Self-Destruction), Probiotics.

* **Corresponding author Kriti Ghatani:** Department of Food Technology, University of North Bengal, Raja Rammohunpur, Darjeeling, West Bengal, 734013, India; E-mail: ghatanik@nbu.ac.in

Prakash M. Halami & Aravind Sundararaman (Eds.)

INTRODUCTION

Bifidobacterium is most abundant in the human gut, as well as other mammalian guts, though other sources were also reported, for example, recently, a *Bifidobacterium* strain, LMG 28769^T, has been isolated from the household water kefir fermentation process [1]. The exact numbers of *Bifidobacterium* species present in the human gut are not well known. It is hypothesized that this genus comprises most of the human gut microbiota. These bacteria are generally anaerobic in nature, rod-like shaped, gram-positive, non-spore-forming, hetero-fermentative, non-motile, catalase-negative, and belong to the phylum *Actinobacteria* [2]. *Bifidobacteria* are mostly high in GC content, and their genomic size generally varies between 1.73 to 3.12Mb. To date, 2500 genome assemblies of *Bifidobacteria* are available in the NCBI database [3]. This bacterium was first isolated in the year 1899 by Henri Tissier from the isolates of newborn infants' fecal matter though it was then classified as *Bacillus bifidus communis,* which was currently termed *Bifidobacterium bifidum* Ti. 16s rRNA. Pangenome analysis is one of the major techniques providing better insight into their genomic component and inter-species relationship as well as their adaptation in a niche. In Table **1**, a few data regarding GC%, amount of tRNA, rRNA, and CRISPR arrays are given for some *Bifidobacterium* species. A comparative table of genome size, genes, and amount of encoded proteins of some *Bifidobacterium* sp. is presented in Fig. (**1**). Several mobilomes (mobile elements) have been found in Bifidobacterial genomes. These consist of IS (Insertion Sequences), plasmids, prophage, and other prophage-like elements. IS*30* is most abundant in *Bifidobacterium* species, though other Insertion Sequences are also found in the different *Bifidobacterium* species like IS *3*, IS *110*, IS *150*, IS *256*, *etc* [4]. IS are found to be important in bacterial adaptation in an environment where these elements are responsible for gene deletion, rearrangement, or other processes for adaptation. There are more than 2000 ORF (Open Reading Frames) per genome found in the *Bifidobacterium* genus by *in silico* gene prediction [5].

However, plasmids are not common in most *Bifidobacteria*, but some of the species carry a significant number of plasmids. For example, pNAC3, pBLO1, pB44, pMB1, *etc.*, plasmids have been found in *B. longum* subsp. *longum* [6 - 8]. Like that, pBC1 and pB80 are found in *B. catenulatum* and *B. bifidum,* respectively [9, 7]. Major replication (Rep) proteins encoded by the Bifidobacterial plasmids are homologous to RCR Rep proteins though some other types, like, homologous to theta replication proteins, are also noticed. There is a lot to further discover in the *Bifidobacterium* genome as it can open new paths for the utilization of these probiotics with improved efficiency and characteristics.

Different species of *Bifidobacteria* have been isolated from different habitats. Mostly they are isolated from human fecal samples, but other sources like animals (like murine, bee, tamarind, marmoset, pig, rabbit, gorilla, chicken, *etc.*), sewage, and food are also reported (shown in Fig. (**2**)), and human origin *Bifidobacterium* species along with their mode of action are listed in Table **2**.

Table 1. Some genome information of different *Bifidobacterium* species.

Bifidobacterium Species	GC%	tRNAs	rRNAs	CRISPR Arrays	Reference
B. angulatum DSM 20098 = JCM 7096	59.4	53	12	1	
B. breve JCM 7017	58.7	53	6	2	
B. bifidum YIT 10347	62.8	52	9	1	
B. animalis subsp. lactis BLC1	60.5	52	12	1	
B. dentium JCM 1195 = DSM 20436	58.5	56	13	2	
B. adolescentis ATCC 15703	59.2	54	16	1	[3]
B. pseudolongum ASM228291v1	63.4	52	12	1	
B. thermophilum RBL67	60.1	47	12	1	
B. pseudocatenulatum YIT11956	56.5	60	16	1	
B. asteroides PRL2011	60.1	45	6	1	

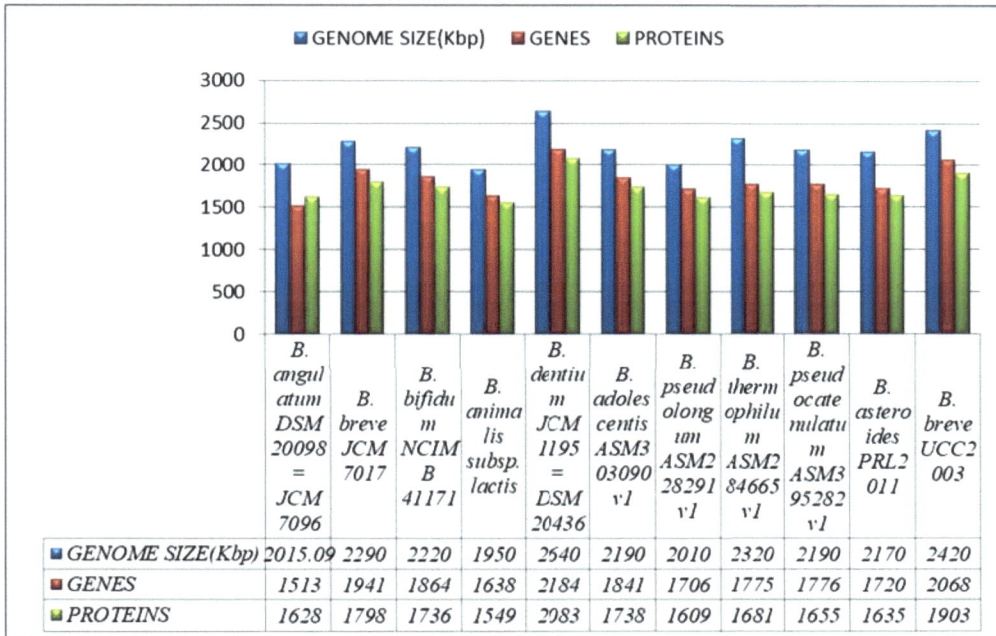

	B. angul atum DSM 20098 = JCM 7096	B. breve JCM 7017	B. bifidu m NCIM B 41171	B. anima lis subsp. lactis	B. dentiu m JCM 1195 = DSM 20436	B. adoles centis ASM3 03090 v1	B. pseud olong um ASM2 28291 v1	B. therm ophilu m ASM2 84665 v1	B. pseud ocate nulatu m ASM3 95282 v1	B. astero ides PRL2 011	B. breve UCC2 003
GENOME SIZE(Kbp)	2015.09	2290	2220	1950	2640	2190	2010	2320	2190	2170	2420
GENES	1513	1941	1864	1638	2184	1841	1706	1775	1776	1720	2068
PROTEINS	1628	1798	1736	1549	2983	1738	1609	1681	1655	1635	1903

Fig. (1). Comparison of genome size, genes, and protein amount of different *Bifidobacterium* species.

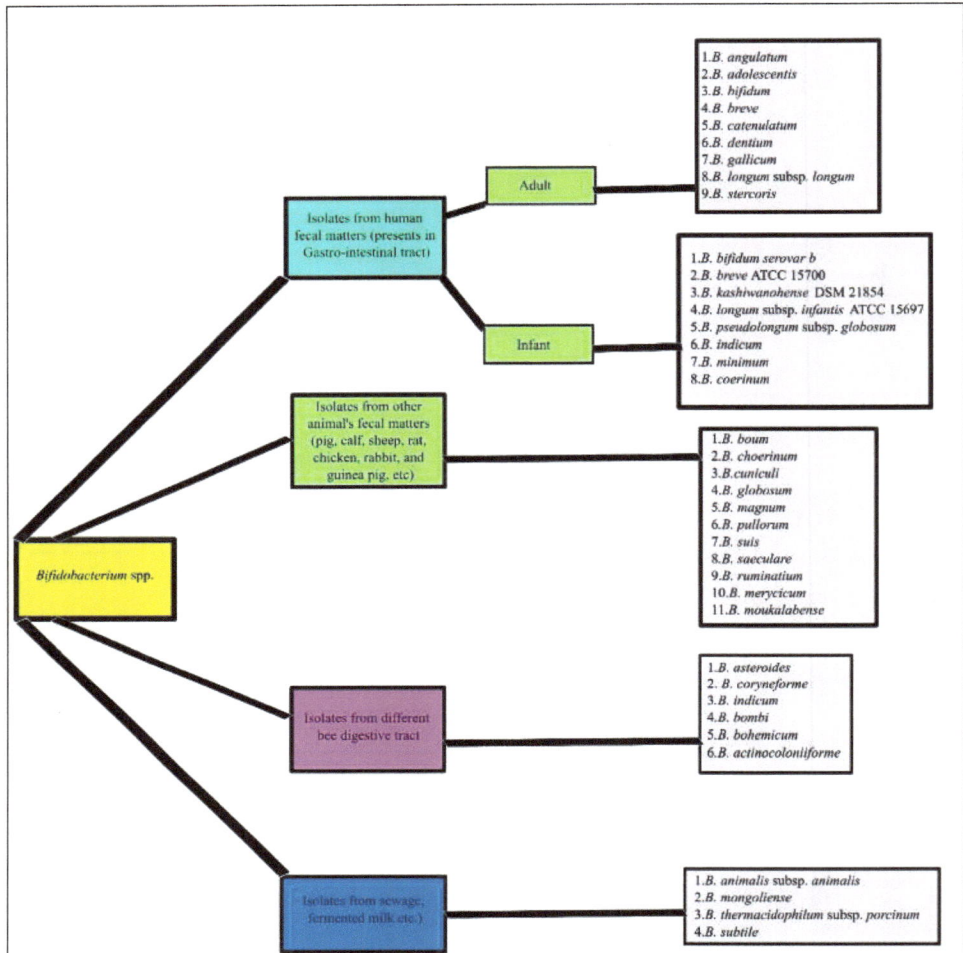

Fig. (2). Different *Bifidobacterium* species and their corresponding isolation sources.

Table 2. Sources of strains identified and their mode of action.

Name of Strain	Sources	Identification Techniques	Mode of Action	References
B. bifidum	Gastro intestinal tract	Quantitative real-time PCR (qPCR)	Ability to metabolize human milk oligosaccharides (HMOs) and have ability to adhere to the intestinal mucosa.	[16]

(Table 2) cont.....

Name of Strain	Sources	Identification Techniques	Mode of Action	References
B. longum	Human gut	Bacterial 16S rRNA gene Pyrosequencing was carried out using a 454 GS FLX Titanium Sequencing System	Anti-inflammatory and inhibit the production of endotoxin by gut microbiota.	[17]
B. animalis subsp. lactis	Human breast milk	16S rDNA gene sequencing using specific primer	Produce exopolysaccharide and stimulates the growth.	[19]
B. breve, B. longum, B. bifidum		Randomly amplified polymorphic DNA (RAPD- PCR) using OPL-16 primer	Show antimicrobial property against enteric pathogen.	[21]
B. breve, B. longum		16S RNA sequencing with specific primer.	Show antimicrobial activity, GIT survival, and adherence to intestinal mucosa.	[25]
B. longum	Human feces	16S rRNA gene sequence with specific primers	Show antagonistic activity against *E. coli* O157:H7.	[26]
B. catenulatum, B. minimum, B. indicum, B. dentium, B. asteroides, B. galicum, B. coerinum. B. indicum	Infant stool	16S rRNA amplification using 27F-1492R primers	Inhibit *E. coli* and *C. albicans*, also possess lysozyme resistance activity, antimicrobial activity, and also produce bacterial exopolysaccharide.	[27]
B. bifidum	Human Vagina	PCR with universal primers and 16S rRNA sequencing	Have ability to adhere to vagina and can eliminate the pathogen by competitive exclusion.	[29]
B. dentium		qPCR with cpn60 amplicon and 16S rRNA sequencing	Increase survival rate of *Lactobacilli* to maintain a healthy vaginal microbiome.	[30]

ORIGIN OF *BIFIDOBACTERIA*

From Human Gut

After birth, the colonisation of pioneer microbes represents the first step in the mutualistic relationship for shaping the developing microbial community [10]. *Bifidobacterium* is one of the most abundant bacterial genera in the healthy infant gut. The abundance of these bacteria decreases with age. During adulthood, the levels of bacteria considerably decrease but remain relatively stable [11]. A different study revealed that initially, colonisation of *Bifidobacteria* occurs from birth. The colonisation may increase and is influenced by several extrinsic factors [12]. It has been observed that from the mother to child, Bifidobacterial species are transmitted vertically and colonise the infant's intestine at the very early stages of life [13]. Evidence suggests that gastrointestinal and immune functions are connected with the brain. Gut microbes produce different metabolites and substances that could affect host brain functions [14]. *Bifidobacterium* is a probiotic bacteria present in the intestinal environment and adheres to the gut epithelium. *Bifidobacterium* has produced surface carbohydrate polymers known as exopolysaccharides (EPS) that act as a protective surface layer and interconnect with the surrounding environment like epithelial mucosa. In fact, some gut microorganisms use this polymer for carbohydrate metabolism and release some bacterial metabolites that exhibit positive benefits for the host [15]. A different study reported that *B. bifidum* is the first coloniser of the newborn's gastrointestinal tract because *B. bifidum* has the ability to metabolize human milk oligosaccharides (HMOs) [16]. *B. longum* (NK46) is isolated from the human gut, which is anti-inflammatory and inhibits the production of endotoxin by gut microbiota [17]. However, in the human gut, the Bifidobacterial sp. inhabits across the whole lifespan. In the harsh environment of the human gastrointestinal tract, these bacteria can survive and give a positive impact on human health [13].

Bifidobacteria Isolated from Breast Milk

During infancy, human breast milk provides critical nutrients and bioactive compounds that support growth and development [18]. Studies suggest that *Bifidobacteria* are present in breast milk. Exopolysaccharide-producing *Bifidobacteria* isolated from human breast milk was observed [19]. *B. animalis* subsp. *lactis*, a probiotic strain isolated from human breast milk has potential against acute and chronic colitis in mice [20]. *Bifidobacterium* includes *B. breve, B. longum,* and *B. bifidum* isolates from human breast milk which show antimicrobial properties against enteric pathogens and have been considered potential probiotic bacteria [21]. Another study reported that *Bifidobacterium* isolates were obtained from breast milk, which acts as potential probiotics and can induce IL-10 from murine macrophages [22]. Probiotic strain *B. lactis* INL1 isolated from human breast milk shows *in vitro* antagonistic activity against gram-positive and gram-negative bacteria by lowering the pH *via* the production of organic acids, and shows the competitive exclusion, the stimulation of the host's immunity, and the production of specific antibacterial substances [23]. Human milk contains complex oligosaccharides (HMOs) which are abundant in milk. However, these complex oligosaccharides are indigestible to infants. Studies have shown that *Bifidobacterium* sp. adapted for utilisation of HMOs act as a probiont and also inhibit the growth of pathogen [24]. Another study reported that three *B. breve* and three *B. longum* strains isolated from human milk show antimicrobial activity, GIT survival, and adherence to intestinal mucosa [25].

From Feces

Another author reported that *B. longum* isolated from the human feces sample showed antagonistic activity against *Escherichia coli* O157:H7. Studies have indicated that *Bifidobacteria* act as a potential probiotic that can protect mice against Shiga toxin-producing *Escherichia coli* O157:H7 lethal infection [26]. Different studies have suggested that *Bifidobacterium* sp. includes *B. catenulatum, B. minimum, B. indicum, B. dentium, B. asteroides, B. galicum,* and *B. coerinum. B. indicum* isolated from the infant stool had the potential to inhibit *E. coli* and *Candida albicans*. These bacterial isolates possess lysozyme resistance activity, antimicrobial activity, and produce exopolysaccharide production [27]. Another study showed that LAB isolated from healthy newborn baby feces acts as a potential probiotic activity such as antibiotic resistance potential. *Bifidobacteria* isolates show antibiotic resistance potential namely erythromycin, gentamicin, oxacillin, ofloxacin, amoxicillin, and cxefotaxime [28].

From Human Vagina

Bifidobacteria is a dominant microbial species present in the vaginal ecosystem together with other lactic acid bacteria. These bacteria can adhere to the vagina and can eliminate the pathogen by competitive exclusion and protect from pathogenic infection. The study reported that *Bifidobacteria* show antibiotic resistance properties against kanamycin, streptomycin, chloramphenicol, gentamicin, and ampicillin [29]. Women's reproductive health depends on the vaginal microbiome and bacterial vaginosis occurs due to imbalances in this microbiota. A study reports that *Bifidobacterium* is the dominant member of some vaginal microbiomes and suggests that *Bifidobacteria* have the potential probiotic bacteria and also increase the survival rate of *Lactobacilli* to maintain a healthy vaginal microbiome [30].

BIFIDOBACTERIUM AS PROBIOTICS

Probiotics are "live microorganisms, when administered in adequate amounts confer a health benefit on the host" [31]. They are non-pathogenic microorganisms, usually of human origin [32]. The core benefits, which are common for all probiotic microorganisms are regulation of intestinal transit, normalization of intestinal microbiota, turnover of enterocytes, competitive exclusion of pathogens, colonization resistance, and short-chain fatty acid production. However, other health benefits of probiotics, which are strain specific are neurological effects, immunological effects, endocrinological effects, and the production of bioactive compounds [33]. Based on many *in vitro* studies and clinical trials, the positive health impact on the consumers of probiotic foods is lactose metabolism, control of gastrointestinal infections, suppression of cancer, reduction of serum cholesterol, and immuno-modulatory activities [34]. Numerous bacterial species belonging to various genus find their use as probiotics. The Lactic acid bacterial members play the most significant role as probiotics. The Lactic acid bacteria used as probiotics for humans include *Lactobacillus acidophilus*, *L. plantarum*, *L. casei*, *L. casei* subsp. *rhamnosus*, *L. delbrueckii* subsp. *bulgaricus*, *L. fermentum*, *L. reuteri*, *Lactococcus lactis* subsp. *lactis*, *Lactococcus lactis* subsp. *cremoris*, *B. bifidum*, *B. infantis*, *B. adolescentis*, *B. longum*, *B. breve*, *Streptococcus salivarius* subsp. *thermophilus*, *Enterococcus faecalis*, *Enterococcus faecium*, *Pediococcus* sp., *etc*. Among these, the *Bifidobacterium* sp. and the *Lactobacillus* sp. are mostly used as probiotics [35].

HEALTH BENEFITS OF *BIFIDOBACTERIA*

A recent upsurge in studies focusing on the Bifidobacterial community and also its use in several commercialized products, especially dairy products like yoghurts, fermented milk, infant formula, *etc.*, understandably reflects its contribution to the well-being of humans [36]. *Bifidobacteria*, a probiotic bacteria can confer numerous health benefits ranging from the improvement of the intestinal microflora composition, improvement of lactose intolerance, anti-mutagenic properties, protection against infections, prevention of rotavirus diarrhoea, traveller's diarrhoea, atopic dermatitis in children, reduction of ulcerative colitis, prevention of necrotizing enterocolitis, prevention of dental caries, *etc* have shown promising reports in many clinical trials [37]. In Table **3** some of their key beneficial activity is given. Recently, exopolysaccharide (EPS) producing strains of *Bifidobacterium* has gathered much attention as EPS contributes to the texture and flavour of the final product and also has immune modulation capability [36, 38]. The changes in the *Bifidobacteria* population in the human microbiomehave been reported to be related to various intestinal and immunological disorders such as irritable bowel syndrome, inflammatory bowel disease (IBD), obesity, and allergy [38]. Although these benefits are strain-dependent, their positive effects cannot be overlooked [36]. The health benefits include the following:

Alleviation of Obesity

Obesity, an increasing health concern around the globe is a combined outcome of genetic and various obesogenic factors. At present, the medications available against obesity are hunger-suppressing drugs and drugs that prevent absorption. However, their uses have been restricted due to their lower efficiency and related side effects. Thus, a better way to tackle this problem is to rely on natural cures. *Bifidobacteria*, the most common probiotic genera and the chief inhabitants of a healthy individual's gut are among the chief natural players that alleviate obesity. The plethora of *Bifidobacteria* in a lean individual suggests that there is a substantial correlation between their abundance and a lean body mass [39]. The observation showed that a higher number of *Bifidobacteria* were present in children with normal body mass in contrast to those children developing obesity [40]. Further, the clinical trials on humans and rats highlight a notable impact of *B. longum* in lowering the total serum cholesterol [41].

Table 3. An overview of clinical trials conducted by various authors using Bifidobacterial strains and their proposed health benefits.

Study Conducted	Probiotic(s)/its Extracts Used	Effects	References
In vivo study on high-fat diet-fed Sprague-Dawley rats by Yin *et al.*, 2010	*Bifidobacterium sp.* (*Bifidobacteria* L66-5, L75-4, M13-4 and FS3-1-1-2)	Reduction of liver and serum triglycerides(TG) and total cholesterol (TCH)	[54]
In vivo study on male rats that were subjected to chronic stress and neurological abnormalities.	*B. breve* strains (CCFM1025 and FHLJDQ3M5)	Anti-depressant-like effect. This may be due to the strain-dependent genomic and metabolic features of *B.breve*.	[55]
In vivo study on mouse with Dextran sodium sulphate (DSS)-induced colitis.	*B. animalis subsp. lactis* strain BB12	Alleviation of DSS-induced colitis thereby suggesting the promising scope of BB12 in the treatment of inflammatory bowel disease.	[56]
In vitro study on colon cancer cell lines (HT-29, HCT- 116, Caco-2) by You *et al.*, 2004.	Chiroinositol containing polysaccharide extracted from *B. bifidum* BGN4	Inhibition of growth of the colon cancer cell lines HT-29 and HCT-116 thereby suggesting the potential of *B.bifidum* in playing an indirect role in inhibiting colon cancer.	[57]
Double blind clinical trial on formula fed infants supplemented with *B. longum sub sp. infantis* strain CECT7210.	*B. longum subsp. infantis* strain CECT7210	This *Bifidobacterium* supplement may reduce the occurrence of diarrhoea in healthy infants.	[58]

Infant Health

Bifidobacteria, the first colonizers of the newborn mammals' intestinal lumen, play a vital role in developing their physiology and maturing the immune system [42]. The early establishment of the *Bifidobacteria* genera in the gut is suggested to prevent necrotizing enterocolitis in infants [43]. *Bifidobacteria* also possesses phosphoprotein phosphatase activity which aids in the breakdown of casein present in the human milk and thereby helps in its absorption in infants [44]. Further, *Bifidobacterium* also possesses a potency of degrading complex sugars. The first-class genes for metabolism are a significant part of the *Bifidobacterium* core genome [45].

Enhancement of Immunity

B. longum, the most abundant member of a breastfed healthy individual has been reported to be capable of preventing the alleviation of colitis [46]. Strain-dependent clinical trials on animals have also suggested the potential of *B. longum* in reducing chronic mucosal inflammation. Pre-clinical studies on *B. animalis* subsp. *lactis* BB-12 suggests its high adhering capability to the intestinal mucosal layer, which further implies that it possesses potent immune-modulatory and pathogen-inhibitory activities [47]. The immuno-modulatory effects studied showed that BB-12 induced dendritic cell maturation and cytokine expression [48]. A notable number of the Bifidobacterial microflora colonize the human colon, however, their number and species composition differ considerably in different individuals. It is suggested that *B. adolescentis* and *B. longum* prevail mostly in adults, while *B. breve* and *B. infantis* are the chief colonizers of the infant gut. *Bifidobacteria*, along with some *Lactobacillus* sp. contribute significantly to the eco-physiology of the colonic microflora. *Bifidobacteria* confer resistance to infections and diarrhoeal diseases [49].

Vitamin Production

The Bifidobacterial community is capable of excreting several water-soluble vitamins, depending on the species and strain. Strains of *B. bifidum* and *B. infantis* are capable of producing substantial amounts of thiamine, nicotinic acid, and folate. *B. breve* also produces pyridoxine and vitamin B12, though they are not excreted [49]. Some strains of *Bifidobacterium* like *B. infantis* CCRC14633 and *B. longum* B6 have been reported to produce B vitamins (riboflavin and thiamine) during the fermentation of soymilk. Riboflavin or thiamine deficiencies can cause liver and skin disorders and alterations in brain glucose metabolism, respectively. Folate (Vitamin B9), which is involved in many crucial processes and hence its deficiency may cause neural tube defects, certain forms of cancer, poor cognitive performance, and coronary heart diseases are synthesized by *Bifidobacteria* during the development of fermented food products. *B. catenulatum* ATCC 27539 has been attributed to the high-level production of folate *in vitro*, while *B. lactis* CSCC5127, *B. infantis* CSCC5187, and *B. breve* CSCC5181 strains have been reported to increase folate concentration during the fermentation of reconstituted skim milk. Cobalamines (Vitamin B12), a common deficient vitamin among vegans are not produced by animals, plants, and fungi. The sole source of its production is microorganisms. It has been studied that cobalamine can be synthesized by *B. animalis* Bb12 in fermented milk. Menaquinone (Vitamin K2), a significant promoter of bone and cardiovascular health is produced by some strains of *Bifidobacterium*. Although menaquinones are ubiquitous in bacteria, some members like Lactobacillus have lost the potential to produce them [38].

Prevention of Constipation

The production of organic acids by *Bifidobacteria* is believed to stimulate the peristaltic movement in the intestine thereby restoring normal bowel movement and helping one get rid of constipation [44].

Anti-bacterial Activity

The carbohydrate metabolism by *Bifidobacteria* yields acetic acid and lactic acid in the molar ratio of 3:2. This yield of acetic acid and lactic acid plays a significant role by controlling the pH in the large intestine and thereby restricting the growth of numerous pathogenic microorganisms [32]. In addition, *Bifidobacteria* is capable of breaking down conjugated bile acids. The breakdown product thus formed *i.e.*, free bile acid aids in the inhibition of pathogens. Furthermore, *in vitro* studies have suggested the anti-bacterial potential of *Bifidobacteria* against pathogenic *E. coli*, *Staphylococcus aureus*, *Shigella dysenteriae*, *Salmonella typhi*, *Proteus* sp. and *C. albicans* [44]. Buffie and Pamer, 2013 have reported the production of organic acids and peptides by some *Bifidobacterium* sp. that has affected the growth and adhesion of *E. coli* [50].

Amino Acid Production

Comparative genome studies of *Bifidobacteria* and *Lactobacillus* throw light on the potential of *Bifidobacterium* to produce at least 19 amino acids and all the enzymes essential for the synthesis of the nucleotides, purine and pyrimidine [45].

Gamma-aminobutyric Acid (GABA) Production

Some, *Bifidobacterium* sp. were also reported to produce gamma-aminobutyric acid (GABA), the chief inhibitory neurotransmitter of the central nervous system, which plays an important role in the induction of hypotension, diuretic effects, anti-diabetic, relaxing, and tranquilizer effects, *etc.* Conjugated linoleic acid (CLA), a polyunsaturated fatty acid (PUFA) biosynthesized by LAB and *Bifidobacteria* through bioconversion of linoleic acid (LA; cis-9, cis-12 C18:2) has been reported to possess anti-carcinogenic, anti-atherogenic, anti-inflammatory, and anti-diabetic activity [38].

Alleviation of Psychological Disorders

Probiotics are an important inhabitant of the gut. Their curative characteristics are very well known to man. Moreover, various kinds of literature have also reinforced the possibility of the relationship between the probiotic microflora of the gut and their influential interplay in the pathogenesis of psychological disorders like depression. Clinical investigations have suggested that the increased concentrations of *Bifidobacteria* and *Lactobacillus* in the gut ameliorate anxiety and temper in patients suffering from chronic fatigue syndrome and inflammatory bowel syndrome [51].

Anti-colorectal Cancer Properties

Colorectal cancer, ranking third among the most common cancer types and a notable reason for death among the population worldwide, correlates to inflammation, microbial exposure, alterations inepigenetics and genetics. Various findings have indicated that the augmentation of microorganisms like *Bifidobacterium, Enterococcus hirae,* and *L. johnsonii* supports the success of platinum-based chemotherapy, immunotherapy, *etc.*, by increasing the anti-tumor immune retaliation [52]. Reports suggest that probiotics like *Bidobacterium* and *Lactobacillus* sp. employ anti-carcinogenic activities by regulating the cell cycle of cancer cells and by making them prone to apoptosis.

Additionally, these species are also capable of producing antioxidative enzymes, chelating heavy metals, and neutralizing various carcinogens. Studies on the anti-cancer properties of Bifidobacterial metabolic secretions in the cell lines of colorectal cancer, HT -29, and Caco-2 show that the metabolites bring about intrinsic and extrinsic apoptotic pathways [53].

BIFIDOBACTERIAL GENOME EDITING

With the advancements in technology, it is now possible to modify the genetic constituents of *Bifidobacteria*. One of them is CRISPR-Cas9, which is quite a new terminology in Biotechnology. In 2020, the Nobel Prize in Chemistry was awarded to Emmanuelle Charpentier and Jennifer A. Doudna for the discovery of the CRISPR-Cas system in the year 2012. However, there are several other processes for the purpose but this one has some huge advantages as compared to the rest. Some methods of *Bifidobacterium* bioengineering are given in Fig. (**3**). The probiotic capability of *Bifidobacteria* species is already well-proven in several types of research and more is going on. But to add this species as a starter culture there are certainly some challenges to overcome and those can only be solved with the help of modern biotechnological tools.

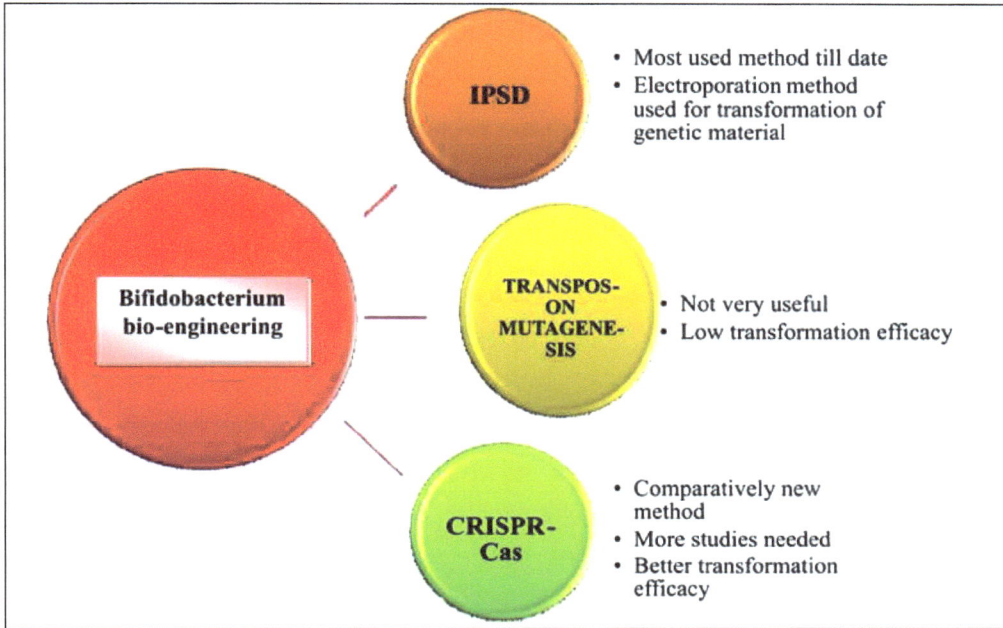

Fig. (3). Different methods for bio-engineering of *Bifidobacterium.*

Bifidobacteria consists of small genomes ranging in size from 2.0-2.8 Mbp [59]. The reason behind the small size of the bifidobacterial genomes can be attributed to the genomic decay (hypothetically) and the potential cause for this can be hypothesized, as the dispensable genes are lost in the process of adaptation in a specific niche. Though the genomic variation among a single species varies hugely among different strains [60], genome editing, an essential tool for improving the beneficial activity of *Bifidobacterium* species, which are one of the earliest and most important colonizers of the human gastrointestinal tract, generally taken part in body's many physiochemical mechanisms as well as shows many symbiotic activities. Philippe and Douady (2003) suggested that Bacteriophages, plasmids, and transposons are the major agents acting in the Horizontal Gene Transfer (HGT) process to modify the Bifidobacterial genomes.

CRISPR-Cas system, the bacterial adaptive defense system helps them stay protected from their predators as these genes are capable of producing specific proteins which act against viral or other predatory attacks. It is evident that the occurrence of the CRISPR-Cas system is not a general characteristic of entire species or subspecies rather it is strain-specific. To date, 2 classes, 6 types, and 34 subtypes have been discovered in this system. Class I system, which consists of types I, III, and IV, has a complex mechanism where several Cas proteins assisted recognition of foreign nucleic acid takes place. Whereas in the Class II system (type II, V, and VI) the single effector protein can perform both recognition and cleavage.

Also, in this class, a type II CRISPR-Cas9 system exists which is unique and important for genome editing [61]. In Table **4**, some of the CRISPR-Cas systems found in *Bifidobacterium* genus are given. The major elements, which are associated with the CRISPR-Cas9 system are crRNA, Cas9 protein, tracrRNA, and PAM. CRISPR-Cas system is quite available in *Bifidobacterium* species thus making it suitable to alter the genetic constituents for improving its stability and probiotic activity. The CRISPR-Cas9 system acts by a system-specific DNA targeting and cleaving method, where the native protein and a duplex complex of crRNA-tracrRNA complete the process. This system has been now well developed for use as a genome editing tool [62]. However, other major probiotic bacteria like *Lactobacillus* sp. have already been subjected to the CRISPR system heavily though *Bifidobacterium* species are the exception.

Table 4. CRISPR-Cas systems in *Bifidobacterium* genus.

Bifidobacterium Species	CRISPR System Type	Cas Protein Type	Reference
B. breve UCC2003	I-C	Cas1,Cas3	
B. dentium LMG 11045	II-C I-C	Cas1, Cas9 Cas1, Cas3	
B. pullorum LMG 21816	I-C	Cas1, Cas3	[63]
B. animalis subsp. *animalis* ATCC 25527	I-E	Cas1, Cas3	
B. animalis subsp. *lactis* DSM 10140	I-U	Cas1, Cas3	
B. bifidum LMG 13200	II-A	Cas1, Cas9	

Inducible Plasmid Self-Destruction (IPSD) can be used for Bifidobacterial genome engineering, which includes gene insertion, deletion and replacement. The *tetW* gene is the reason behind the tetracycline resistance capability of major dominant Bifidobacterial species found in human gastrointestinal tracts. It might be horizontally transferred to other species, so to prevent the horizontal transfer of this gene and to keep the safety of probiotics it is better to delete the functional antibiotic-resistant genes. An IPSD plasmid pBIZrec-ΔtetW was successfully used to inactivate the *tetW* gene by an in-frame deletion in *B. longum* IF3-53, some infant feces isolate which shows tetracycline resistance and increased the tetracycline sensitivity among the mutant strain (1 µg/ml than from 32 µg/ml of parent strain) [64].

Another work provides insight into how capable the IPSD vectors really are, it is chromosomal insertion and expression of heterologous genes in *Bifidobacterium*. Catalase encoding gene *LpKatL* isolated from *L. plantarum* was inserted into the chromosome of *B. longum* NCC 2705 under the control of P$_{hup}$ promoter. As per researchers, the recombinant strain possesses catalase activity in the presence of hematin. It also shows improved viability when exposed to H$_2$O$_2$ as compared to the parent strain. IPSD plasmid can be used successfully for genome editing in *Bifidobacteria*. IPSD strategy can be well utilized in a wide range of synthetic biology applications of *Bifidobacteria* species mostly in the food and pharmaceutical industry. Using Cre-*lox* site-specific recombination system an IPSD plasmid, pBIZrec which was constructed from pDP870, carried the spectinomycin resistance gene. This plasmid was later introduced into *B. longum* NCC 2705 and thus the strain having the resistance against spectinomycin antibiotic. The *L. monocytogenes* containing the *BetL* gene consistently increase its ability of osmotolerance, barotolerance, and acid tolerance. Therefore, the introduction of this gene fragment (*BetL*) into the probiotic strain of *B. breve* UCC2003 can significantly increase its ability to tolerate gastric acids and elevated osmolarity and will be very useful [65].

B. subtilis contains the *KatE* gene, which encodes for catalases. An expression vector pBCAT001 was derived from plasmid pKKT427 (by insertion of *KatE* gene of *B. subtilis* within the promoter and terminator of *hup* of *B. longum* and introduced into *B. longum* 105-A in Fig. (**4**). It helps to improve the stress tolerance capability of *B. longum* 105-A and helps in the accumulation of H$_2$O$_2$ [66]. Guglielmetti *et al.* (2008) tried to produce a Bifidobacterial biosensor to analyze the physiological state of cells under different environmental conditions by introducing a plasmid vector pGBL8b containing luciferase gene of a click beetle (*Pyrophorus plagiophthalamus*) into *B. longum* biovar *longum*. There are many other aspects where engineered *Bifidobacteria* using the IPS technique can be used.

Ulcerative Colitis, a kind of chronic Inflammatory Bowel Disease (IBD) that generally occurs in the rectum and colon, is often described as a chronic inflammation in the colonic mucosal lining caused by several factors such as oxidative stress, triggered immune system against the good bacterial colony that takes part in body's many physiochemical and gastrointestinal processes, *etc* [67]. However, to date, no exact medications are available to treat this disease, but several novel approaches have been made and one such is oral supplementation of the fusion protein by *B. longum* species. In research, *E. coli* DH5α and the *B. longum* HB15 were used as a cloning host and a transformation host respectively [67]. pBDMnSOD plasmid vector was made by incorporating oligonucleotide containing hMnSOD and other gene fragments into the pBV220 vector. The recombinant *B. longum*-PEP-1-rhMnSOD was made by electroporating the pBDMnSOD into the *B. longum* HB15 strain [67]. This particular strain was then administered into the dextran sulfate sodium (DSS)-induced ulcerative colitis mice by the researchers and they observed surprising results such as in this case there were reduced inflammatory cytokines TNF-α, IL-1βand IL-6 levels as well as increased anti-inflammatory cytokine IL-8 levels. Thus, it can be concluded that the probiotic *Bifidobacterium* strain can be used to treat DSS-induced ulcerative colitis [67].

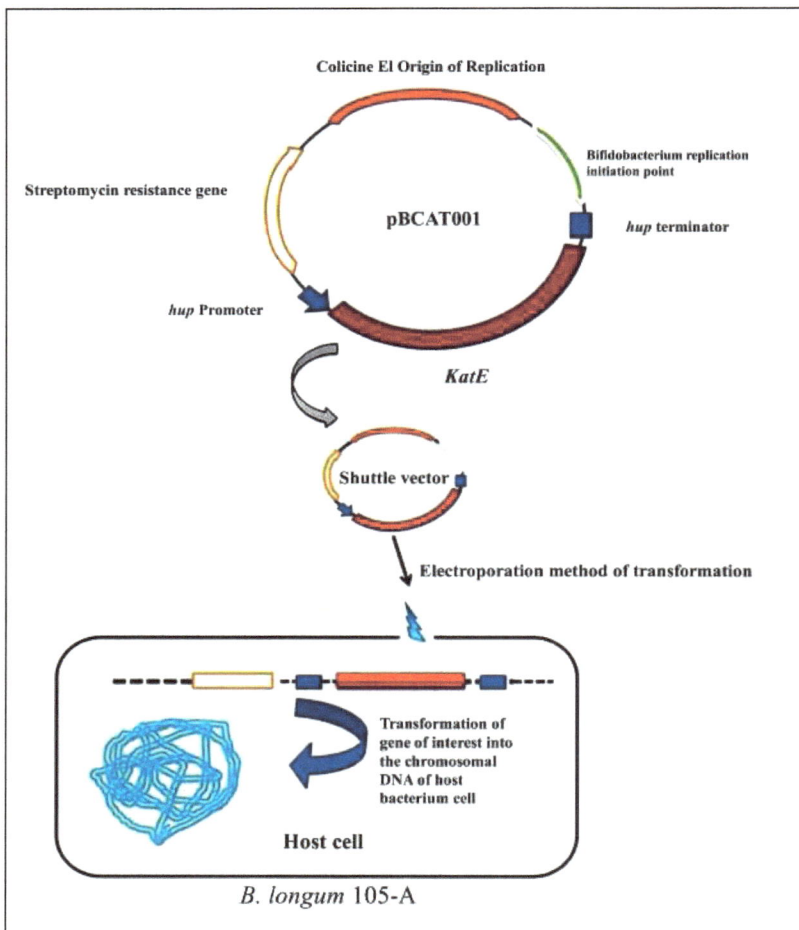

Fig. (4). Modified plasmid pBCT001 which contains *KatE* gene, streptomycin resistance gene, *Bifidobacterium* replication initiation site, the origin of replication site, *hup* terminator, and *hup* promoter region, transferred into host *B. longum* 105-A *via* electroporation method. This later expresses the gene of interest in the host cell.

In a recently published study, by the use of double crossover recombination techniques the *B. animalis* subsp. *lactis* DSM10140 was mutated by non-synchronous mutation. Both the original and the mutated strain labeled with two fluorescent proteins m-Cherry and Green Fluorescent Protein by Tu promoter was successfully used to study the role of EPS in the adhesion and biofilm production comparatively [68].

It is important for any probiotics to have the ability of different stress tolerance and *Bifidobacteria* are mostly sensitive to stress tolerance. Oxidative stress is a major problem for the *Bifidobacterium* species, which must be overcome to successfully use these as probiotics. Several genome editing approaches have been made to overcome this problem. As reported earlier, *B. longum* has the ability to tolerate oxidative stress though that is very little. The alkyl hydroperoxide reductase catalytic subunit C (ahpC) is the reason behind their ability, which take part in various biochemical pathways to convert the various reactive oxygen species (ROS) to other non-harmful chemicals. Therefore, researchers tried to over-express the gene fragment in the strain to see the problem. They have isolated the ahpC from *B. longum* and amplified it in PCR. The pDP401-ahpC containing the isolated gene fragment was transformed into the *B. longum* NCC2705 by electroporation method. The recombinant *B. longum* NCC2705 possesses many characteristics; most importantly, it facilitates several pathways to eliminate ROS species [69]. In another research, *LpKatL* and *StSodA* genes were isolated from *L. plantarum* strain NL42 and *S. thermophilus* strain XJ41 chromosomal DNA respectively, which codes for heme-dependent catalase and superoxide dismutase (SOD) respectively. These two chemicals are already well studied for the roles of their ROS-scavenging pathways and have proven to help many anaerobic microbes tolerate oxidative stress. The isolated genes are then incorporated by expression plasmids (pDP-Kat-Sod) into *B. longum* NCC2705. The resulting recombinant strain expressed by both enzymes is proven to reduce oxidative damage by intra and extra-cellular H_2O_2, moreover helps in tolerating intense oxidative stress by various synergistic effects [69]. Thus, these strains can be well utilized in various oxidative stress conditions. As per another research heat shock proteins (hsp) can successively increase the different stress tolerance abilities. The *hsp20* gene fragment which encodes for sHsp, is PCR amplified and transferred into the pMD19-T plasmid then transferred into the *E. coli* DH5α. Recombinant plasmid pDP401-PsHsp and the control plasmid pDP401 were constructed from that and inserted into *B. longum* NCC2705 electro-competent cells by electroporation method. This recombinant strain is able to tolerate high amounts of heat and salt, which can be very useful in producing effective probiotics as well as starter cultures in dairy industries [70].

In a study, researchers have tried to develop a novel method for drug delivery system by editing the genome of probiotic strain *B. longum* NCC2705. The expression vector pBBAD-Tum was used to deliver the tumstatin, a powerful angiostatin coding gene fragment into the strain by electroporation techniques. At the *Bpi*I and *Xba*I sites of the pBBAD-GFP plasmid, the tumstatin encoding gene *Tum* was used to replace the *GFP* (Green Fluorescent Protein) gene for the construction of pBBAD-Tum. When administered to the tumor-induced mice, they showed impressive results, like anti-tumor effects, apoptotic vascular endothelial cells higher number counts [71]. They proposed that it could be a good method for solid tumor therapy.

A novel method for developing oral vaccines by the use of genetically modified *Bifidobacteria* is currently in trend. Enterotoxigenic *E. coli* is mostly responsible for diarrhea. The recombinant plasmids, pBEX-Cfab, and pBEX-LTB are used to express antigens responsible for ETEC in the gut which triggers long-term immune response against these. The fimbrial subunit (Cfab) and *eltB* (gene encodes for LTB) of ETEC extracted from H10407 plasmid and ETEC E44815 respectively, were used to construct pBEX-Cfab, pBEX-LTB respectively. These plasmids were then transferred into *B. infantis*. When administered into the host it gives a good result [72]. In the future, *Bifidobacterium* based vaccination can play a vital role.

B. breve UCC2003 was subjected to genome editing for the expression of human basic fibroblast growth factor (FGF-2) which has several biological functions like cell migration, cell proliferation, cell differentiation, angiogenesis, *etc* [7]. *B. breve* UCC2003 *sec2* signal peptide coding sequences were used along with *B. longum* VMKB44 *gap* promoter and *hup* terminator for producing the vector plasmid. In the pESH45 *sec2,* signal peptide coding sequences were inserted along with the synthetic human *FGF-2* gene containing pλFGFB (*Nde*I/*Bam*HI fragment). The expression vector based on pESH80 (pESH86) was then constructed by transferring the gene of interest of recombinant pESH45 plasmid into pESH80 and the following was then transferred into the *B. breve* UCC2003 by electroporation method for the successful expression of HGF-2 [7]. This recombinant strain successfully expressed the particular protein of interest as the researchers tested the amount of this by ELISA and other methods. Thus, they concluded that it could be possible to administer the vaccine or drug through these probiotic strains. In Table **5** some common plasmids, which are used for successive protein expression in different *Bifidobacterium* species have been given.

Table 5. Plasmids used in genetic alteration and their expressed recombinant protein in host *Bifidobacterium* species.

Bifidobacteria Species	Plasmid	Protein Expressed	References
B. pseudocatenulatum	pMDYP469AbfB	*B. longum* arabinofuranosidase	[6]
B. breve	pLuxMC2	*Photorhabdus luminescens* luciferase	[73]
	pESH93	Human interleukin 10	[74]
	pESH100	Human interleukin 10	[74]
	pLuxMC3	*P. luminescens* luciferase	[73]
	pMDYP469AbfB	*B. longum* arabinofuranosidase	[6]
	pAV001-HU-eCD	*P. luminescens* luciferase	[75]
	pBIFRIBO-gusA	β-Glucuronidase	[76]
	pESH86	Human fibroblast growth factor	[77]
B. longum	pBES16PR-CHOL	*Streptomyces coelicolor* cholesterol oxidase	[78]
	pPSAB1	*Pediococcus* sp. pediocin PA-1	[79]
	Pgusa	*E. coli* β –glucuronidase	[80]
	pLR2	Synthetic human interleukin 10	[81]
	pGBL8b	Firefly luciferase	[82]
	pKKT427	B. subtilis catalases (heme dependent)	[66]
	pBBADs-OXM	Human oxyntomodulin	[83]
	pBBADs-IL-12	Human interleukin 12	[7]
	pBES2	α-amylase of *B. adolescentis*	[84]
	pAV001-HU-eCD-M968	*E. coli* cytosine deaminase	[85]
	pJW245	*Salmonella typhimurium* FliC	[86]
	pBV220/endostatin	Human endostatin	[87]
B. adolescentis	pMDYP469AbfB	*B. longum* arabinofuranosidase	[6]

Zinc Finger Nucleases (ZFN), which mostly comprise two components: i) DNA binding Zinc Finger domain and ii) DNA cleavage domain, are able to cleave specific target DNA. Transcription Activator-Like Effector Nucleases (TALEN) is another type of genome editing technique, which has also two components: i) TAL effector DNA-binding domain and ii) DNA cleavage domain. This system works by the activity of Transcription activator-like effectors to bind the particularly targeted DNA strands and the DNA cleavage domain, which contains endonucleases, cleaves the strand. In addition, Homology Directed Repair can successively introduce foreign DNA at the cleaved site. However, several new methodologies have been developed but it cannot be said all of them are well developed to be used in the bioengineering of *Bifidobacterium*. ZFN and TALEN are not well studied for the *Bifidobacterium* but several studies are going on for the potential use of these methods in finding new paths.

CONCLUSION

Genome editing is a modern technique by which genetic material can be added altered or removed at a specific location in the genome. Different genome editing techniques such as CRISPR-Cas, IPSD, ZFN, and TALEN are used for the development of genetically engineered organisms. Genome editing is nowadays used for the prevention and treatment of human disease. In this chapter, we focused on the Bifidobacterial genome editing technique by which we can accelerate the production of different compounds such as bioactive peptides, enzymes, and some effector molecules to increase the product's nutritional quality and increase the desired health benefit. The CRISPR-Cas9 technique is one of the most useful techniques to modify genetic constituents of *Bifidobacteria* to improve the beneficial property for the development of strain-specific probiotic attributes. Alteration and modification of the Bifidobacterial genome can increase the positive health benefit of the production of amino acid and vitamin, gamma-aminobutyric acid, the chief inhibitory neurotransmitter of the central nervous system having anti-diabetic, relaxing, and tranquilizer effects and these bacteria can prevent disease such as diarrhoea, prevention of colorectal cancer, reduction of ulcerative colitis. CRISPR-Cas9 is a widely used technique used for Bifidobacterial genome editing where TALEN and ZFN are not well studied for *Bifidobacterium* but several studies are going on for potential use of these methods in finding new paths. In this chapter, we presented a comprehensive study of Bifidobacterial genome editing tools and their application in the food industry as a potential probiotic bacterium in dairy and non-dairy starter cultures for the development of various fermented foods.

ACKNOWLEDGEMENTS

The authors would like to thank the Department of Food Technology, University of North Bengal, for providing the computational infrastructure and library facility.

REFERENCES

[1]   Laureys D, Cnockaert M, De Vuyst L, Vandamme P. *Bifidobacterium aquikefiri* sp. nov., isolated from water kefir. Int J Syst Evol Microbiol 2016; 66(3): 1281-6.
[http://dx.doi.org/10.1099/ijsem.0.000877] [PMID: 26739269]

[2]   Ventura M, van Sinderen D, Fitzgerald GF, Zink R. Insights into the taxonomy, genetics and physiology of *Bifidobacteria*. Antonie van Leeuwenhoek 2004; 86(3): 205-23.
[http://dx.doi.org/10.1023/B:ANTO.0000047930.11029.ec] [PMID: 15539925]

[3] *Bifidobacterium* - NLM [Internet]. National Center for Biotechnology Information. U.S. National Library of Medicine Available from: https://www.ncbi.nlm.nih.gov/search/all/?term=bifidobacterium

[4] Lee JH, O'Sullivan DJ. Genomic insights into *Bifidobacteria*. Microbiol Mol Biol Rev 2010; 74(3): 378-416.
[http://dx.doi.org/10.1128/MMBR.00004-10] [PMID: 20805404]

[5] Makarova K, Slesarev A, Wolf Y, *et al.* Comparative genomics of the lactic acid bacteria. Proc Natl Acad Sci 2006; 103(42): 15611-6.
[http://dx.doi.org/10.1073/pnas.0607117103] [PMID: 17030793]

[6] Corneau N, Émond É, LaPointe G. Molecular characterization of three plasmids from *Bifidobacterium longum*. Plasmid 2004; 51(2): 87-100.
[http://dx.doi.org/10.1016/j.plasmid.2003.12.003] [PMID: 15003705]

[7] Shkoporov AN, Efimov BA, Khokhlova EV, Kafarskaia LI, Smeianov VV. Production of human basic fibroblast growth factor (FGF-2) in *Bifidobacterium breve* using a series of novel expression/secretion vectors. Biotechnol Lett 2008; 30(11): 1983-8.
[http://dx.doi.org/10.1007/s10529-008-9772-8] [PMID: 18575808]

[8] Matteuzzi D, Brigidi P, Rossi M, Di D. Characterization and molecular cloning of *Bifidobacterium longum* cryptic plasmid pMB1. Lett Appl Microbiol 1990; 11(4): 220-3.
[http://dx.doi.org/10.1111/j.1472-765X.1990.tb00165.x] [PMID: 1366860]

[9] Álvarez-Martín P, O'Connell-Motherway M, van Sinderen D, Mayo B. Functional analysis of the pBC1 replicon from *Bifidobacterium catenulatum* L48. Appl Microbiol Biotechnol 2007; 76(6): 1395-402.
[http://dx.doi.org/10.1007/s00253-007-1115-5] [PMID: 17704917]

[10] Lawson MAE, O'Neill IJ, Kujawska M, *et al.* Breast milk-derived human milk oligosaccharides promote *Bifidobacterium* interactions within a single ecosystem. ISME J 2020; 14(2): 635-48.
[http://dx.doi.org/10.1038/s41396-019-0553-2] [PMID: 31740752]

[11] Odamaki T, Kato K, Sugahara H, *et al.* Age-related changes in gut microbiota composition from newborn to centenarian: A cross-sectional study. BMC Microbiol 2016; 16(1): 90.
[http://dx.doi.org/10.1186/s12866-016-0708-5] [PMID: 27220822]

[12] Arboleya S, Watkins C, Stanton C, Ross RP. Gut *Bifidobacteria* populations in human health and aging. Front Microbiol 2016; 7: 1204.
[http://dx.doi.org/10.3389/fmicb.2016.01204] [PMID: 27594848]

[13] Wong CB, Odamaki T, Xiao J. Insights into the reason of human-residential *Bifidobacteria* (HRB) being the natural inhabitants of the human gut and their potential health-promoting benefits. FEMS Microbiol Rev 2020; 44(3): 369-85.
[http://dx.doi.org/10.1093/femsre/fuaa010] [PMID: 32319522]

[14] Kiraly DD, Walker DM, Calipari ES, *et al.* Alterations of the host microbiome affect behavioral responses to cocaine. Sci Rep 2016; 6(1): 35455.
[http://dx.doi.org/10.1038/srep35455] [PMID: 27752130]

[15] Castro-Bravo N, Wells JM, Margolles A, Ruas-Madiedo P. Interactions of surface exopolysaccharides from *Bifidobacterium* and *Lactobacillus* within the intestinal environment. Front Microbiol 2018; 9: 2426.
[http://dx.doi.org/10.3389/fmicb.2018.02426] [PMID: 30364185]

[16] Duranti S, Lugli GA, Milani C, *et al.* *Bifidobacterium bifidum* and the infant gut microbiota: An intriguing case of microbe-host co-evolution. Environ Microbiol 2019; 21(10): 3683-95.
[http://dx.doi.org/10.1111/1462-2920.14705] [PMID: 31172651]

[17] Lee HJ, Lee KE, Kim JK, Kim DH. Suppression of gut dysbiosis by *Bifidobacterium longum* alleviates cognitive decline in 5XFAD transgenic and aged mice. Sci Rep 2019; 9(1): 11814.
[http://dx.doi.org/10.1038/s41598-019-48342-7] [PMID: 31413350]

[18] Lyons KE, Ryan CA, Dempsey EM, Ross RP, Stanton C. Breast milk, a source of beneficial microbes and associated benefits for infant health. Nutrients 2020; 12(4): 1039.
[http://dx.doi.org/10.3390/nu12041039] [PMID: 32283875]

[19] Kansandee W, Moonmangmee D, Moonmangmee S, Itsaranuwat P. Characterization and *Bifidobacterium sp.* growth stimulation of exopolysaccharide produced by *Enterococcus faecalis* EJRM152 isolated from human breast milk. Carbohydr Polym 2019; 206: 102-9.
[http://dx.doi.org/10.1016/j.carbpol.2018.10.117] [PMID: 30553302]

[20] Burns P, Alard J, Hrdỳ J, *et al.* Spray-drying process preserves the protective capacity of a breast milk-derived *Bifidobacterium lactis* strain on acute and chronic colitis in mice. Sci Rep 2017; 7(1): 43211.
[http://dx.doi.org/10.1038/srep43211] [PMID: 28233848]

[21] Eshaghi M, Bibalan MH, Rohani M, *et al.* *Bifidobacterium* obtained from mother's milk and their infant stool; A comparative genotyping and antibacterial analysis. Microb Pathog 2017; 111: 94-8.
[http://dx.doi.org/10.1016/j.micpath.2017.08.014] [PMID: 28826763]

[22] Oddi S, Binetti A, Burns P, *et al.* Occurrence of bacteria with technological and probiotic potential in argentinian human breast-milk. Benef Microbes 2020; 11(7): 685-702.
[http://dx.doi.org/10.3920/BM2020.0054] [PMID: 33161735]

[23] Zacarías MF, Souza TC, Zaburlín N, *et al.* Influence of technological treatments on the functionality of *Bifidobacterium lactis* INL1, a breast milk-derived probiotic. J Food Sci 2017; 82(10): 2462-70.
[http://dx.doi.org/10.1111/1750-3841.13852] [PMID: 28892139]

[24] Bidart GN, Rodríguez-Díaz J, Monedero V, Yebra MJ. A unique gene cluster for the utilization of the mucosal and human milk-associated glycans galacto- N -biose and lacto- N -biose in *Lactobacillus casei*. Mol Microbiol 2014; 93(3): 521-38.
[http://dx.doi.org/10.1111/mmi.12678] [PMID: 24942885]

[25] Solís G, de los Reyes-Gavilan CG, Fernández N, Margolles A, Gueimonde M. Establishment and development of lactic acid bacteria and *Bifidobacteria* microbiota in breast-milk and the infant gut. Anaerobe 2010; 16(3): 307-10.
[http://dx.doi.org/10.1016/j.anaerobe.2010.02.004] [PMID: 20176122]

[26] Inoue H, Shibata S, Ii K, Inoue J, Fukuda S, Arakawa K. Complete genome sequence of *Bifidobacterium longum* strain jih1, isolated from human feces. Microbiol Resour Announc 2020; 9(22): e00319-20.
[http://dx.doi.org/10.1128/MRA.00319-20] [PMID: 32467270]

[27] Kusharyati DF, Rovik A, Ryandini D, *et al.* *Bifidobacterium* from infant stool: The diversity and potential screening. Biodiversitas 2020; 21(6).
[http://dx.doi.org/10.13057/biodiv/d210623]

[28] Lase E, Davidson AL, Lister INE, Fachrial E. Probiotic activity and antibiotic sensitivity of lactic acid bacteria isolated from healthy breastfed newborn baby feces. IOP Conference Series: Materials Science and Engineering 2021; 1071(1): 12015.
[http://dx.doi.org/10.1088/1757-899X/1071/1/012015]

[29] Sirichoat A, Flórez AB, Vázquez L, *et al.* Antibiotic susceptibility profiles of lactic acid bacteria from the human vagina and genetic basis of acquired resistances. Int J Mol Sci 2020; 21(7): 2594.
[http://dx.doi.org/10.3390/ijms21072594] [PMID: 32276519]

[30] Freitas AC, Hill JE. Quantification, isolation and characterization of *Bifidobacterium* from the vaginal microbiomes of reproductive aged women. Anaerobe 2017; 47: 145-56.
[http://dx.doi.org/10.1016/j.anaerobe.2017.05.012] [PMID: 28552417]

[31] Hotel AC, Cordoba A. Health and nutritional properties of probiotics in food including powder milk with live lactic acid bacteria. Prev Sci 2001; 5(1): 1-0.
[PMID: 11519371]

[32] Ashraf R, Shah NP. Selective and differential enumerations of *Lactobacillus delbrueckii subsp. bulgaricus, Streptococcus thermophilus, Lactobacillus acidophilus, Lactobacillus casei* and *Bifidobacterium spp.* in yoghurt — A review. Int J Food Microbiol 2011; 149(3): 194-208.
[http://dx.doi.org/10.1016/j.ijfoodmicro.2011.07.008] [PMID: 21807435]

[33] Soni R, Jain NK, Shah V, Soni J, Suthar D, Gohel P. Development of probiotic yogurt: Effect of strain combination on nutritional, rheological, organoleptic and probiotic properties. J Food Sci Technol 2020; 57(6): 2038-50.
[http://dx.doi.org/10.1007/s13197-020-04238-3] [PMID: 32431330]

[34] Charalampopoulos D, Wang R, Pandiella SS, Webb C. Application of cereals and cereal components in functional foods: A review. Int J Food Microbiol 2002; 79(1-2): 131-41.
[http://dx.doi.org/10.1016/S0168-1605(02)00187-3] [PMID: 12382693]

[35] Conway PL. Selection criteria for probiotic microorganisms. Asia Pac J Clin Nutr 1996; 5(1): 10-4.
[PMID: 24394458]

[36] Wong CB, Sugahara H, Odamaki T, Xiao JZ. Different physiological properties of human-residential and non-human-residential *Bifidobacteria* in human health. Benef Microbes 2018; 9(1): 111-22.
[http://dx.doi.org/10.3920/BM2017.0031] [PMID: 28969444]

[37] Langa S, Peirotén A, Gaya P, *et al.* Human *Bifidobacterium* strains as adjunct cultures in Spanish sheep milk cheese. J Dairy Sci 2020; 103(9): 7695-706.
[http://dx.doi.org/10.3168/jds.2020-18203] [PMID: 32684453]

[38] Linares DM, Gómez C, Renes E, *et al.* Lactic acid bacteria and *Bifidobacteria* with potential to design natural biofunctional health-promoting dairy foods. Front Microbiol 2017; 8: 846.
[http://dx.doi.org/10.3389/fmicb.2017.00846] [PMID: 28572792]

[39] Ray M, Hor PK, Ojha D, Soren JP, Singh SN, Mondal KC. *Bifidobacteria* and its rice fermented products on diet induced obese mice: Analysis of physical status, serum profile and gene expressions. Benef Microbes 2018; 9(3): 441-52.
[http://dx.doi.org/10.3920/BM2017.0056] [PMID: 29409330]

[40] Kalliomäki M, Carmen Collado M, Salminen S, Isolauri E. Early differences in fecal microbiota composition in children may predict overweight. Am J Clin Nutr 2008; 87(3): 534-8.
[http://dx.doi.org/10.1093/ajcn/87.3.534] [PMID: 18326589]

[41] Xiao JZ, Kondo S, Takahashi N, *et al.* Effects of milk products fermented by *Bifidobacterium longum* on blood lipids in rats and healthy adult male volunteers. J Dairy Sci 2003; 86(7): 2452-61.
[http://dx.doi.org/10.3168/jds.S0022-0302(03)73839-9] [PMID: 12906063]

[42] Speranza B, Bevilacqua A, Campaniello D, *et al.* The impact of gluten friendly flour on the functionality of an active drink: Viability of *Lactobacillus acidophilus* in a fermented milk. Front Microbiol 2018; 9: 2042.
[http://dx.doi.org/10.3389/fmicb.2018.02042] [PMID: 30214438]

[43] Gaucher F, Bonnassie S, Rabah H, *et al.* Review: adaptation of beneficial *Propionibacteria, Lactobacilli,* and *Bifidobacteria* improves tolerance toward technological and digestive stresses. Front Microbiol 2019; 10: 841.
[http://dx.doi.org/10.3389/fmicb.2019.00841] [PMID: 31068918]

[44] Arunachalam KD. Role of *Bifidobacteria* in nutrition, medicine and technology. Food Nutr Res 1999; 19(10): 1559-97.

[45] Lukjancenko O, Ussery DW, Wassenaar TM. Comparative genomics of *Bifidobacterium, Lactobacillus* and related probiotic genera. Microb Ecol 2012; 63(3): 651-73.
[http://dx.doi.org/10.1007/s00248-011-9948-y] [PMID: 22031452]

[46] Yan S, Yang B, Zhao J, *et al.* A ropy exopolysaccharide producing strain *Bifidobacterium longum* subsp. *longum* YS108R alleviates DSS-induced colitis by maintenance of the mucosal barrier and gut microbiota modulation. Food Funct 2019; 10(3): 1595-608.

[http://dx.doi.org/10.1039/C9FO00014C] [PMID: 30806428]

[47] Flach J, Van der Waal MB, Kardinaal AFM, Schloesser J, Ruijschop RMAJ, Claassen E. Probiotic research priorities for the healthy adult population: A review on the health benefits of *Lactobacillus rhamnosus* GG and *Bifidobacterium animalis* subspecies *lactis* BB-12. Cogent Food Agric 2018; 4(1): 1452839.
[http://dx.doi.org/10.1080/23311932.2018.1452839]

[48] Latvala S, Pietilä TE, Veckman V, *et al.* Potentially probiotic bacteria induce efficient maturation but differential cytokine production in human monocyte-derived dendritic cells. World J Gastroenterol 2008; 14(36): 5570-83.
[http://dx.doi.org/10.3748/wjg.14.5570] [PMID: 18810777]

[49] Macfarlane GT, Steed H, Macfarlane S. Bacterial metabolism and health-related effects of galacto-oligosaccharides and other prebiotics. J Appl Microbiol 2008; 104(2): 305-44.
[PMID: 18215222]

[50] Long RT, Zeng WS, Chen LY, *et al.* *Bifidobacterium* as an oral delivery carrier of oxyntomodulin for obesity therapy: inhibitory effects on food intake and body weight in overweight mice. Int J Obes 2010; 34(4): 712-9.
[http://dx.doi.org/10.1038/ijo.2009.277] [PMID: 20065960]

[51] Desbonnet L, Garrett L, Clarke G, Kiely B, Cryan JF, Dinan TG. Effects of the probiotic *Bifidobacterium infantis* in the maternal separation model of depression. Neuroscience 2010; 170(4): 1179-88.
[http://dx.doi.org/10.1016/j.neuroscience.2010.08.005] [PMID: 20696216]

[52] Yoon Y, Kim G, Jeon BN, Fang S, Park H. *Bifidobacterium* strain-specific enhances the efficacy of cancer therapeutics in tumor-bearing mice. Cancers 2021; 13(5): 957.
[http://dx.doi.org/10.3390/cancers13050957] [PMID: 33668827]

[53] Faghfoori Z, Faghfoori MH, Saber A, Izadi A, Yari Khosroushahi A. Anticancer effects of *Bifidobacteria* on colon cancer cell lines. Cancer Cell Int 2021; 21(1): 258.
[http://dx.doi.org/10.1186/s12935-021-01971-3] [PMID: 33397383]

[54] Yin YN, Yu QF, Fu N, Liu XW, Lu FG. Effects of four *Bifidobacteria* on obesity in high-fat diet induced rats. World J Gastroenterol 2010; 16(27): 3394-401.
[http://dx.doi.org/10.3748/wjg.v16.i27.3394] [PMID: 20632441]

[55] Tian P, Bastiaanssen TFS, Song L, *et al.* Unraveling the microbial mechanisms underlying the psychobiotic potential of a *Bifidobacterium breve* Strain. Mol Nutr Food Res 2021; 65(8): 2000704.
[http://dx.doi.org/10.1002/mnfr.202000704] [PMID: 33594816]

[56] Chae JM, Heo W, Cho HT, *et al.* Effects of orally-administered *Bifidobacterium animalis* subsp. *lactis* strain BB12 on dextran sodium sulfate-induced colitis in mice. J Microbiol Biotechnol 2018; 28(11): 1800-5.
[http://dx.doi.org/10.4014/jmb.1805.05072] [PMID: 30270609]

[57] You HJ, Oh DK, Ji GE. Anticancerogenic effect of a novel chiroinositol-containing polysaccharide from *Bifidobacterium bifidum* BGN4. FEMS Microbiol lettrs 2004; 240(2): 131-6.

[58] Escribano J, Ferré N, Gispert-Llaurado M, *et al.* *Bifidobacterium longum* subsp *infantis* CECT7210-supplemented formula reduces diarrhea in healthy infants: a randomized controlled trial. Pediatr Res 2018; 83(6): 1120-8.
[http://dx.doi.org/10.1038/pr.2018.34] [PMID: 29538368]

[59] Bottacini F, Medini D, Pavesi A, *et al.* Comparative genomics of the genus *Bifidobacterium*. Microbiology 2010; 156(11): 3243-54.
[http://dx.doi.org/10.1099/mic.0.039545-0] [PMID: 20634238]

[60] Milani C, Lugli GA, Duranti S, *et al.* Genomic encyclopedia of type strains of the genus *Bifidobacterium*. In: Griffiths MW, Lugli GA, Duranti S, Eds. Appl Environ Microbiol. 2014; 80: pp.

(20)6290-302.

[61] Jinek M, Chylinski K, Fonfara I, Hauer M, Doudna JA, Charpentier E. A programmable dual-RN-
-guided DNA endonuclease in adaptive bacterial immunity. Science 2012; 337(6096): 816-21.
[http://dx.doi.org/10.1126/science.1225829] [PMID: 22745249]

[62] Cong L, Ran FA, Cox D, *et al.* Multiplex genome engineering using CRISPR/Cas systems. Science
2013; 339(6121): 819-23.
[http://dx.doi.org/10.1126/science.1231143] [PMID: 23287718]

[63] Briner AE, Lugli GA, Milani C, *et al.* Occurrence and diversity of CRISPR-Cas systems in the genus
Bifidobacterium. PLoS One 2015; 10(7): e0133661.
[http://dx.doi.org/10.1371/journal.pone.0133661] [PMID: 26230606]

[64] Zuo F, Zeng Z, Hammarström L, Marcotte H. Inducible Plasmid Self-Destruction (IPSD) assisted
genome engineering in *Lactobacilli* and *Bifidobacteria.* Bio Rxiv. 2019; pp. 1-12.

[65] Sheehan VM, Sleator RD, Hill C, Fitzgerald GF. Improving gastric transit, gastrointestinal persistence
and therapeutic efficacy of the probiotic strain *Bifidobacterium breve* UCC2003. Microbiology
(Reading) 2007; 153(10): 3563-71.
[http://dx.doi.org/10.1099/mic.0.2007/006510-0] [PMID: 17906153]

[66] He J, Sakaguchi K, Suzuki T. Acquired tolerance to oxidative stress in *Bifidobacterium longum* 105-A
via expression of a catalase gene. Appl Environ Microbiol 2012; 78(8): 2988-90.
[http://dx.doi.org/10.1128/AEM.07093-11] [PMID: 22307289]

[67] Liu M, Li S, Zhang Q, Xu Z, Wang J, Sun H. Oral engineered *Bifidobacterium longum* expressing
rhMnSOD to suppress experimental colitis. Int Immunopharmacol 2018; 57: 25-32.
[http://dx.doi.org/10.1016/j.intimp.2018.02.004] [PMID: 29455070]

[68] Castro-Bravo N, Hidalgo-Cantabrana C, Rodriguez-Carvajal MA, Ruas-Madiedo P, Margolles A.
Gene replacement and fluorescent labeling to study the functional role of exopolysaccharides in
Bifidobacterium animalis subsp. *lactis.* Front Microbiol 2017; 8: 1405.
[http://dx.doi.org/10.3389/fmicb.2017.01405] [PMID: 28790996]

[69] Zuo F, Yu R, Khaskheli GB, *et al.* Homologous overexpression of alkyl hydroperoxide reductase
subunit C (ahpC) protects *Bifidobacterium longum* strain NCC2705 from oxidative stress. Res
Microbiol 2014; 165(7): 581-9.
[http://dx.doi.org/10.1016/j.resmic.2014.05.040] [PMID: 24953679]

[70] Fukiya S, Hirayama Y, Sakanaka M, Kano Y, Yokota A. Technological advances in bifidobacterial
molecular genetics: application to functional genomics and medical treatments. Biosci Microbiota
Food Health 2012; 31(2): 15-25.
[http://dx.doi.org/10.12938/bmfh.31.15] [PMID: 24936345]

[71] Wei C, Xun AY, Wei XX, *et al. Bifidobacteria* expressing tumstatin protein for antitumor therapy in
tumor-bearing mice. Technol Cancer Res Treat 2016; 15(3): 498-508.
[http://dx.doi.org/10.1177/1533034615581977] [PMID: 25969440]

[72] Ma Y, Luo Y, Huang X, Song F, Liu G. Construction of *Bifidobacterium infantis* as a live oral vaccine
that expresses antigens of the major fimbrial subunit (CfaB) and the B subunit of heat-labile
enterotoxin (LTB) from enterotoxigenic *Escherichia coli.* Microbiology (Reading) 2012; 158(2): 498-
504.
[http://dx.doi.org/10.1099/mic.0.049932-0] [PMID: 22053005]

[73] Cronin M, Knobel M, O'Connell-Motherway M, Fitzgerald GF, Van Sinderen D. Molecular dissection
of a bifidobacterial replicon. Appl Environ Microbiol 2007; 73(24): 7858-66.
[http://dx.doi.org/10.1128/AEM.01630-07] [PMID: 17965208]

[74] Hidaka A, Hamaji Y, Sasaki T, Taniguchi S, Fujimori M. Taniguchi Si, Fujimori M. Exogeneous
cytosine deaminase gene expression in *Bifidobacterium breve* I-53-8w for tumor-targeting
enzyme/prodrug therapy. Biosci Biotechnol Biochem 2007; 71(12): 2921-6.

[http://dx.doi.org/10.1271/bbb.70284] [PMID: 18159091]

[75] Gill SR, Pop M, DeBoy RT, *et al.* Metagenomic analysis of the human distal gut microbiome. Science 2006; 312(5778): 1355-9.
[http://dx.doi.org/10.1126/science.1124234] [PMID: 16741115]

[76] Pokusaeva K, Neves AR, Zomer A, *et al.* Ribose utilization by the human commensal *Bifidobacterium breve* UCC2003. Microb Biotechnol 2010; 3(3): 311-23.
[http://dx.doi.org/10.1111/j.1751-7915.2009.00152.x] [PMID: 21255330]

[77] Powell IB, Achen MG, Hillier AJ, Davidson BE. A simple and rapid method for genetic transformation of lactic *Streptococci* by electroporation. Appl Environ Microbiol 1988; 54(3): 655-60.
[http://dx.doi.org/10.1128/aem.54.3.655-660.1988] [PMID: 16347576]

[78] Monk IR, Gahan CGM, Hill C. Tools for functional postgenomic analysis of *Listeria monocytogenes.* Appl Environ Microbiol 2008; 74(13): 3921-34.
[http://dx.doi.org/10.1128/AEM.00314-08] [PMID: 18441118]

[79] Long RT, Zeng WS, Chen LY, *et al. Bifidobacterium* as an oral delivery carrier of oxyntomodulin for obesity therapy: Inhibitory effects on food intake and body weight in overweight mice. Int J Obes 2010; 34(4): 712-9.
[http://dx.doi.org/10.1038/ijo.2009.277] [PMID: 20065960]

[80] Iwata M, Morishita T. The presence of plasmids in *Bifidobacterium breve.* Lett Appl Microbiol 1989; 9(5): 165-8.
[http://dx.doi.org/10.1111/j.1472-765X.1989.tb00315.x]

[81] Reyes Escogido ML, De León Rodríguez A, Barba de la Rosa AP. A novel binary expression vector for production of human IL-10 in *Escherichia coli* and *Bifidobacterium longum.* Biotechnol Lett 2007; 29(8): 1249-53.
[http://dx.doi.org/10.1007/s10529-007-9376-8] [PMID: 17487549]

[82] Fu GF, Li X, Hou YY, Fan YR, Liu WH, Xu GX. *Bifidobacterium longum* as an oral delivery system of endostatin for gene therapy on solid liver cancer. Cancer Gene Ther 2005; 12(2): 133-40.
[http://dx.doi.org/10.1038/sj.cgt.7700758] [PMID: 15565182]

[83] Klijn A, Moine D, Delley M, Mercenier A, Arigoni F, Pridmore RD. Construction of a reporter vector for the analysis of *Bifidobacterium longum* promoters. Appl Environ Microbiol 2006; 72(11): 7401-5.
[http://dx.doi.org/10.1128/AEM.01611-06] [PMID: 16997985]

[84] Rhim SL, Park MS, Ji GE. Expression and secretion of *Bifidobacterium adolescentis* amylase by *Bifidobacterium longum.* Biotechnol Lett 2006; 28(3): 163-8.
[http://dx.doi.org/10.1007/s10529-005-5330-9] [PMID: 16489493]

[85] Fukuda S, Toh H, Hase K, *et al. Bifidobacteria* can protect from enteropathogenic infection through production of acetate. Nature 2011; 469(7331): 543-7.
[http://dx.doi.org/10.1038/nature09646] [PMID: 21270894]

[86] Ruiz L, O'Connell-Motherway M, Zomer A, De los Reyes-Gavilán CG, Margolles A, Van Sinderen D. A bile-inducible membrane protein mediates bifidobacterial bile resistance. Microb Biotechnol 2012; 5(4): 523-35.
[http://dx.doi.org/10.1111/j.1751-7915.2011.00329.x] [PMID: 22296641]

[87] Xu Y-F, Zhu L-P, Hu B, *et al.* A new expression plasmid in *Bifidobacterium longum* as a delivery system of endostatin for cancer gene therapy. Cancer Gene Ther 2007; 14(2): 151-7.
[http://dx.doi.org/10.1038/sj.cgt.7701003] [PMID: 17068487]

CHAPTER 5

Metabolic Engineering of *Bifidobacterium* sp. Using Genome Editing Techniques

Aravind Sundararaman[1] and **Prakash M. Halami**[1,*]

[1] Department of Microbiology and Fermentation Technology, CSIR- Central Food Technological Research Institute, Mysuru-570020, India

Abstract: The gut microbiome is significant in maintaining human health by facilitating absorption and digestion in the intestine. Probiotics have diverse and significant applications in the health sector, so probiotic strains require an understanding of the genome-level organizations. Probiotics elucidate various functional parameters that control their metabolic functions. Gut dysbiosis leads to inflammatory bowel disease and other neurological disorders. The application of probiotic bacteria to modulate the gut microbiota prevents diseases and has gained large interest. In a recent decade, the development of modern tools in molecular biology has led to the discovery of genome engineering. Synthetic biology approaches provide information about diverse biosynthetic pathways and also facilitate novel metabolic engineering approaches for probiotic strain improvement. The techniques enable engineering probiotics with the desired functionalities to benefit human health. This chapter describes the recent advances in probiotic strain improvement for diagnostic and therapeutic applications *via* CRISPR-Cas tools. Also, the application of probiotics, current challenges, and future perspectives in disease treatment are discussed.

Keywords: Genome editing, Metabolic engineering, Probiotics, Strain improvement.

INTRODUCTION

The human gastrointestinal tract harbors complex, diverse microbes that regulate the host's physiological functions and well-being. The gut microbiota is linked with many functions, including absorption and fermentation of complex carbohydrates, developing immune functions, and inhibiting pathogen adherence to the intestinal cell wall [1]. The probiotics inhabiting the gastrointestinal tract are influenced by several factors, such as breastfeeding, mode of delivery, age, geography, gender, long-term dietary intake, and drug usage [2].

---

[*] **Corresponding author Prakash M. Halami:** Department of Microbiology and Fermentation Technology, CSIR-Central Food Technological Research Institute, Mysuru-570020, India; E-mail: prakashalami@cftri.res.in

Many diseases, such as diabetes, coronary heart disease, inflammatory bowel disease, and obesity, observed dysbiosis in the gut microbiota composition [3]. Thus, gut microbiota is a complex inheritable trait determined by genetic and environmental factors [4].

Bifidobacterium sp. was first described decades ago and is linked with healthy intestinal tracts due to their abundance in breastfed infants. They are considered probiotics as they elicit health benefits to the host by maintaining intestinal microbial balance and have led to wide application as probiotic components, especially in fermented dairy products. *Bifidobacterium* sp. is a Gram-positive, anaerobic, and saccharolytic bacteria found inhabiting the human oral cavity, and the intestinal tract of mammals [5]. The identification and classification of *Bifidobacterium* species are based on DNA-DNA hybridization and whole genome or conserved sequence phylogenetic analysis [6]. *Bifidobacteria* exert their biological activities by producing antimicrobial substances and vitamins. They also have anti-inflammatory, and anti-obesity properties and regulate the immune system [7].

The genus *Bifidobacterium* encodes genes associated with a broad range of non-digestible carbohydrate utilization, including plant fiber and human milk oligosaccharides [8]. *Bifidobacteria* utilize the hexose sugars through the "bifid shunt" pathway with the help of the vital enzyme fructose-6-phosphate phosphoketolase [9]. The ATP-generating pathway produces short-chain fatty acids (SCFAs) that act as antagonists against pathogens and protect from infections [10]. For example, acetate produced by *Bifidobacteria* protects the host against lethal infections and improves intestinal defense mediated by epithelial cells [11].

Several species of *Bifidobacteria*, such as *B. longum, B. breve*, and *B. animalis*, are used to treat inflammatory bowel disease and gastrointestinal disorders [12]. A few strains, such as *B. animalis subsp. lactis* BB-12 and *B. animalis* BF052, are used as major ingredients in probiotic product formulation, indicating that probiotic characteristics are strain-dependent [13]. Therefore, understanding the metabolism of *Bifidobacteria* completely and its adaptation to various nutrient environments is essential.

In the recent decade, efforts have been aimed at understanding the gut microbiota using metabolic modeling, which generates testable hypotheses to elucidate each species' metabolism and interspecies metabolic interactions [14]. Recently, Thiele and co-workers [15] developed a resource AGORA (Assembly of Gut Organisms through Reconstruction and Analysis) to study the human gut microbiome at the genome scale that enables the system-level study. The genome-scale metabolic models of gut microbes predict the growth phenotypes and link dietary intake and absorption in humans [16]. Further, studies on metabolic reconstruction provide insights into strain-specific metabolic networks and the applications of different strains. In this chapter, we describe the significance of *Bifidobacteria* as a probiotic and their ability to produce metabolites. Moreover, applications of genome engineering in *Bifidobacteria* to improve the production of metabolites to improve human health are discussed.

TYPES OF METABOLITES PRODUCED BY *BIFIDOBACTERIA*

The genera *Bifidobacterium* and *Lactobacillus* are the primary source of probiotics, defined as live microbial food ingredients that exert beneficial effects on health. They are recognized to prevent or treat infections in the large and small intestines. Recent research has demonstrated their antagonistic effect on *H. pylori* diseases [17]. Probiotics have been reported to produce organic acids, immunomodulatory properties, and competitive inhibition for mucous, the binding sites of mucous cells to exert an antagonistic effect on *H. pylori* [18]. The potential uses of probiotic *Bifidobacteria* to combat *H. pylori* infections are less exploited than those reported for *Lactobacilli*.

The antimicrobial metabolites secreted by *Bifidobacteria* are a possible alternative to antibiotic treatment against clinical infections and serve as an adjunct therapy along with simultaneous antibiotic administration (Fig. **1**). The antimicrobial peptides isolated from *Bifidobacteria* could have the prophylactic potential to treat *H. pylori* infections.

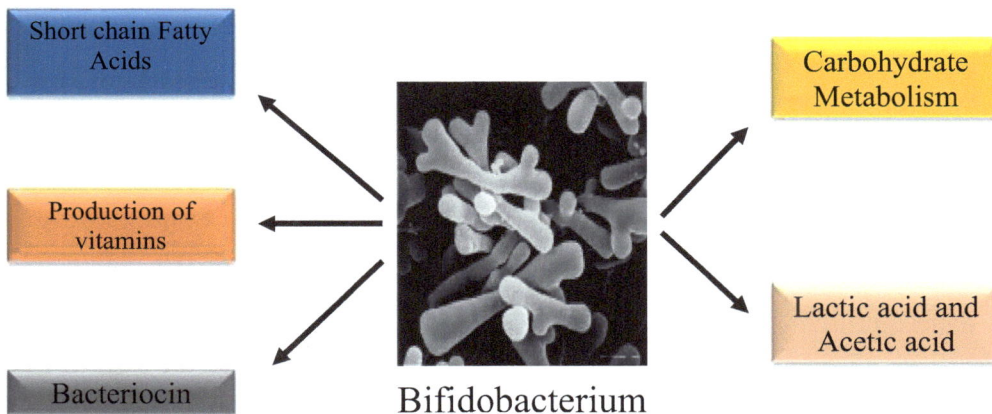

Fig. (1). Potential properties of *Bifidobacterium* used in various applications for genome editing.

Bifidobacteria are classified into three distinct groups based on metabolic capabilities for diverse applications:

1. The *B. bifidum* NCIMB 41171 and BGN4 strains are well delineated from other strains based on their SCFA production, phenotypic data, and differences among the bifid shunt pathway.
2. The strains of *B. animalis* represent high similarity to *B. longum* and *B. thermacidophilum subsp. thermacidophilum* DSM 15837 in terms of phenotypic and metabolic characteristics during the applications in the food industry.
3. The *B. longum* strains of the same species are classified into different groups due to the variations in metabolic and nutrient capabilities.

The genomic sequence analysis of *Bifidobacteria* indicates the presence of at least 19 amino acid biosynthetic genes and encodes enzymes responsible for the biosynthesis of pyrimidine and purine nucleotides, as well as for vitamin B [19].

Vitamin Metabolism

The genome analysis of *Bifidobacteria* revealed to produce a range of Vitamin B, including thiamine (B1), riboflavin (B2), nicotinic acid (or niacin; B3), pyridoxine (B6), and folic acid (B11) [20]. The bifidobacterial strains have been explored for probiotic properties due to their vitamin-producing ability. For example, screening of folate-producing *Bifidobacteria* to provide proliferating colonocytes with folic acid was evaluated in nine different species. The ingestion of folate-producing *Bifidobacteria* prevents folate deficiency in intestinal epithelia and protects against inflammation and cancer [21].

Sugar Metabolism

The complex polysaccharides transit through the digestive tract to the colon and stimulate the proliferation of probiotics. The competitiveness of probiotics is determined by the ability of bacteria to metabolize complex carbohydrates in the GIT [22]. *Bifidobacteria* are observed to catabolize complex carbohydrates. The functional characterization of the *B. dentium* Bd1 genome revealed that 14% of genes are involved in carbohydrate and transport metabolism. The increased availability of different energy sources in the oral cavity may potentiate the highest percentage of gene diversity in the species [23].

Bifidobacteria are saccharolytic organisms that ferment glucose, fructose, and galactose. However, the fermentation of carbohydrates and alcohol differs among species and strains. The physiological properties of *Bifidobacteria* show that they can utilize a wide range of carbohydrates such as fructooligosaccharides (FOS), pectin, xylooligosaccharides (XOS), and mucin. However, glycosyl hydrolases remain unidentified [24]. The genome of *B. breve* UCC2003 has been reported to have *apuB* enzyme, responsible for complex polysaccharides such as starch [25]. *B. bifidum* PRL2010 has shown acquired metabolic ability from host-derived glycans, which indicates the genus *Bifidobacterium* has diverse carbohydrate metabolism.

Pro-, Pre-and Syn-Biotic Concepts

Research on the metabolism of *Bifidobacteria* has led to the synbiotic and prebiotic concepts. A prebiotic is defined as "a selectively fermented ingredient that allows specific changes, both in the composition and activity of the gastrointestinal microflora, that confers benefits upon host well-being and health." Prebiotic oligosaccharides also confer health benefits through anti-adhesive activity. Specifically, the prebiotic oligosaccharides directly inhibit infections by mimicking the pathogen binding site that is present as a coating to the epithelial cells.

FOS and GOS are the most commonly used prebiotics in the European and Japanese markets. However, isomalto-oligosaccharides (IMO), Soybean oligosaccharides (SOS), and xylooligosaccharides (XOS) are not exploited commercially. In addition, the functional analysis of sugar breakdown abilities in *Bifidobacteria* revealed many candidate prebiotic compounds. For example, the genome of *B. breve* UCC2003 identified two novel α-glucosidases, which indicates honey-derived oligosaccharides may be the potential prebiotics for this strain [25].

The prebiotic potential to impact human health is explored by analyzing human milk components and identifying that neutral oligosaccharides in human milk are the likely prebiotic factors that stimulate *Bifidobacteria* growth in the infant's gut. These oligosaccharide constituents have a significant role in the innate immune system where infants are protected from pathogen colonization through breastfeeding. The applications of prebiotics are vital for the development of novel therapeutic and prophylactic agents to treat mucosal pathogens [26].

The prophylactic treatment includes supplementation of symbiotic microbes and a combination of prebiotics and probiotics. The potential benefits of symbiotic supplementation are that they increase the activity of probiotics and effective delivery in the gut. A symbiotic combination of *B. lactis* and resistant starch has demonstrated significant protection against colorectal cancer in rat models [27], indicating that combination therapy seems a superior strategy over pre- or probiotic alone.

Bacteriocins Produced by *Bifidobacterium* sp.

Bifidobacterium sp. are significant symbiotic bacteria of the human intestine accounting for 91% of microorganisms in newborns and 3–7% of the total microorganisms in adults. The species are known to produce both acetic and lactic acid through their metabolic process. The acidic environment formed by *Bifidobacterium* sp. inhibits pathogen colonization and positively affects host health. In addition, *Bifidobacterium* sp. such as *Bifidobacterium bifidum*, *B. longum*, *B. infantum*, and *B. thermophilus* have shown to produce bacteriocins that facilitate pathogen inhibition in the intestine.

Bifidin

Bifidin, a thermostable bacteriocin produced by *B. bifidum* 1452 is stable at 100 °C for 30 minutes. The bifidin can be extracted by ethanol-acetone precipitation, and the extract is stable at 5–8 °C, for three months. The amino acid sequence analysis of the antimicrobial peptide shows the presence of phenylalanine and glutamic acid, and trace amounts of aspartate, glycine, leucine, isoleucine, serine, and threonine. The antimicrobial activity of bifidin is confirmed by *in vitro* experiments against the propagation of *Micrococcus flavus* and *Staphylococcus aureus*.

The metabolites of *B. infantum* BCRC 14602 have been shown to produce the antibacterial substance Bifidin I. It can inhibit Gram-positive and Gram-negative pathogens such as *Bacillus, Streptococcus*, *Staphylococcus, Shigella*, and *Escherichia coli,* respectively. However, it cannot inhibit yeast. The relative molecular weight is 30,000 Da. and the enzyme activity of 1600 AU/mL is obtained after 16–20 h of fermentation in MRS medium. It is stable at pH 4–10.

Bifilong

Bifilong is produced by *B. longum* Bb-46 and found to inhibit *Bacillus cereus*, *E. coli*, *Staphylococcus aureus*, and *Salmonella typhimurium*. The bacteriocin is sensitive to pepsin and trypsin and liable to heat. However, Bifilong is stable between pH 4 and 7 and the activity decreases rapidly above pH 9.

Bifidocin B

Bifidocin B, belonging to class IIa bacteriocin, is the first isolated bacteriocin from *B. bifidum* NCFB 1454 with a molecular weight of 33,000 Da. They are resistant to a pH range of 2-10, thermostable up to 121 °C for 15 mins, and organic solvents, but sensitive to protease. Bifidocin B exhibits antimicrobial properties against foodborne pathogens such as *Bacillus, Enterococcus, Listeria*,

Leuconostoccus, and *Phanerococcus*. The maximum yield (3,200 AU/mL) of Bifidocin B is produced at the exponential phase of growth.

Thermophilicin B67

The strain *B. thermophilum* RBL67 isolated from infant feces produces thermophilicin B67. In simulated conditions of proximal colon, the addition of *B. thermophilus RBL67* resulted in a sharp decrease in the number of *Bifidobacterium* count. However, there is no change in the activity of intestinal flora and microbial composition. The result may be due to the antimicrobial effect of *B. thermophilus* RBL67 producing thermophilicin B67 [28].

Pediococcus acidilactici UVA1, a bacteriocin-producing bacterium isolated from infant feces, when mixed with *B. thermophilus* RBL67, the culture system was stable with bacteriocin activity, and cell growth was not affected. *B. thermophilus* RBL67 mixed with prebiotic oligosaccharides and galactooligosaccharides in simulated conditions of pig intestinal tract, inhibits the growth of *Salmonella enteric* subsp. *enterica* serotype N-15 [29].

The *Clostridium difficile*-associated disease (CDAD) model is used to determine the effects of *B. longum* ATCC 15707 on infection with *C. difficile*. Before injecting the mice with *C. difficile*, an antibiotic cocktail (kanamycin, gentamicin, colistin, metronidazole, and vancomycin) disrupted the intestinal microbiota. Subsequently, *B. longum* ATCC 15707 was administered in mice to evaluate the effect of probiotics. All infected mice showed symptoms of CDAD, including diarrhea, hunching posture, and weight loss, and became moribund in 2-3 days.

The *B. longum* ATCC 15707 treated CDI mice show a higher survival rate (30% mortality) than the untreated CDI mice which showed 64% mortality. Probiotics have gained interest as a potential therapeutic modality for the prevention and treatment of CDAD. Moreover, various studies have reported that probiotics improve gut health by inhibiting pathogen colonization in the gut [30]. Probiotics also produce antimicrobial compounds, including short-chain fatty acids and bacteriocins, that suppress the growth of pathogenic bacteria by increasing the host's immunity through their cell wall components. *B. longum* ATCC 15707 produced organic acid, which led to the lowering of pH and, consequently, growth-inhibition of *C. difficile* when cocultured [31].

GENOME ENGINEERING OF METABOLITES

To improve human health, there is a gaining interest in traditional probiotics such as *Lactobacilli* and *Bifidobacteria* towards developing living diagnostics and therapeutics. However, the mechanism underlying their inherent health beneficial properties is not well investigated, which can delay the application of probiotics in pharmaceutical and functional food formulation. Genome editing techniques elucidate the physiological properties, and molecular mechanism of probiotics and their interaction with the host microbiota. Further, the development of next-generation probiotics with CRISPR-Cas and IPSD tools promotes improved robustness and tailored functionalities of probiotics. The traditional genome editing in *Bifidobacteria* included the applications of plasmid-mediated homologous recombination which were non-replicative and temperature sensitive [32, 33]. The efficiency of the strategies relies on transformation efficiency and has disadvantages such as non-seamless editing, unstable mutations, and limited host range [34]. In addition, counter-selection systems based on *pyrE* (encoding orotate phosphoribosyltransferase) and *upp* (encoding uracil phosphoribosyltransferase) genes facilitate the selection of double-crossover events. However, the parental strains need modification to improve the strategy [35].

Recently, the applications of the CRISPR-Cas system are harnessed in probiotics including *Bifidobacteria* which is a flexible and universal strategy (Table **1**). CRISPR-Cas in probiotics was first applied in *Lactobacillus reuteri* as a counter-selection system to assist ssDNA-mediated recombineering [36]. The single-strand breaks on the chromosomal DNA were generated by CRISPR-Cas9D10A nickase that triggers homology-directed repair (HDR) for precise and rapid chromosomal manipulation by the one-step transformation in *Lactobacillus casei* [37]. However, sometimes heterologous Cas nuclease may be toxic to bacterial cells.

Table 1. Genome editing technologies applied for probiotics including *Bifidobacteria*.

Species/Strains	Technology	Benefits	References
Bifidobacterium breve	Non-replicative plasmid-mediated homologous recombination	Single transformation step	[38]
Bifidobacterium longum	Temperature-sensitive plasmid-mediated homologous recombination	Stable mutation; Seamless editing	[35]
Lactobacillus casei *Lactobacillus plantarum*	Prophage recombinase-assisted dsDNA or ssDNA recombineering	Stable mutation; Seamless editing	[37]
Lactobacillus reuteri *Lactobacillus crispatus*	CRISPR-Cas system-based genome editing	Transformation efficiency-independent; Seamless editing; High efficiency; Stable mutation	[36]

(Table 1) cont.....

Species/Strains	Technology	Benefits	References
Lactobacillus gasseri *Bifidobacterium* longum	Inducible Plasmid Self-Destruction (IPSD) assisted genome engineering	Transformation efficiency-independent; Stable mutation; Seamless editing; Flexible and universal tool for both *Lactobacilli* and *Bifidobacteria*	[34]

Genome editing in *Lactobacillus crispatus* has been achieved through endogenous type I CRISPR-Cas system through plasmid transformation. *Bifidobacteria* are reported to have an abundant CRISPR-Cas system [39], which may be exploited for genome editing. However, genome editing with the CRISPR-Cas tools has significant barriers, such as strain-specific, that need to be addressed for efficient editing and broader application in different species [40].

Strategies to Improve the Transformation Efficiency

The genome engineering tools are effective, although their application depends on the transformation of respective plasmids. The transformation conditions in lactic acid bacteria have been recently reviewed by Wang *et al.* [41]. The factors affecting electroporation include characteristics of the host strains, parameters of electric pulse, treatment of cell wall weakening agent, and recovery medium. In *Bifidobacteria*, NaCl treatment and prolonged recovery cultivation contribute to improving the transformation efficiency. Moreover, cultivation of *Lactiplantibacillus plantarum* (previously *Lactobacillus plantarum*) recipient cells at 21°C increases transformation efficiency.

Restriction-modification (RM) systems are the primary genetic barrier to DNA transfer and act as bacterial defense mechanisms against invading foreign DNA. A strategy called plasmid artificial modification (PAM) can bypass the RM system through *in vitro* methylation of plasmid prior to transformation. The PAM strategy improves transformation efficiency by 2-5 magnitude and is a predominant method for transformation in *Bifidobacteria* [42]. However, the laborious construction of PAM recombinants and the diversity of the RM system in the bacterial host make the method difficult to rapidly apply in genome editing. An alternative strategy is to inactivate genes encoding restriction endonucleases in the parental strains and perform electroporation in DNA restriction system-deficient strains. However, this may lead to an increased susceptibility of the strains to bacteriophages. Plasmid methylation of the recipient cells to bypass the RM system, greatly simplifies the complicated procedures in the PAM strategy.

Inducible Plasmid Self-Destruction (IPSD) Assisted Genome Engineering

Inducible Plasmid Self-Destruction (IPSD) assisted genome engineering was developed as a tool for a flexible and universal strategy based on replicative plasmid [34]. The tool can be used to perform gene insertion, deletion, and replacement in *Bifidobacteria* to elucidate the molecular mechanisms of probiotics in host cell adaptation [43]. For the application of IPSD, the replicon, and the antibiotic resistance gene are separated by site-specific recombinase-targeted DNA fragments from two orientations, and an inducible promoter drives the corresponding site-specific recombinase gene. Once the expression is induced, the plasmid may recombine between the targeted DNA fragments at site-specific recombinase and lose function due to the excision of the replicon.

The plasmid delivery of homologous DNA into bacteria initiates natural recombination and IPSD facilitates the selection of single-crossover integrated precursor and homologous recombination. In the absence of antibiotics, the single crossover mutants can be obtained and the double-crossover mutant could be obtained by selecting antibiotic-sensitive colonies. The IPSD method could be a potential approach to engineer various bacteria since it does not depend on transformation efficiency.

Study on Probiotic Bacteria-host Interactions by Genome Editing

The gold standard approach to understanding the gene function relationship is mutagenesis, especially in microorganisms [44]. In *Lactobacilli*, bacteria-host interactions have been studied by the construction of gene knockout mutants. For *Lactobacilli*, the strategies based on conditional replicative plasmid coupled with a nonessential gene, *upp*-based counter selection system is developed. The system has been used to study the role of cell surface components involved in stress tolerance, cell morphology, and host-pathogen interactions of probiotics. CRISPR–Cas9-assisted ssDNA recombineering [36] has been used to investigate phage production in *L. reuteri*. The study revealed IPSD-assisted genome engineering for activation of the Ack pathway *via* fructose metabolism and environmental short-chain fatty acids (SCFAs) combination in *L. reuteri* modulates phage production in the gut. The strategy helps identify the adhesion factors in vaginal probiotics [43]. The important role of exopolysaccharides (EPS) and a novel sortase-dependent protein in the adhesion of *Lactobacillus gasseri* DSM 14869 to vaginal epithelial cells have been studied by producing knockout mutants that seemed to confer a specific niche adaptation trait to the probiotics. The specific role of sortase-dependent proteins (SDPs) in the adhesion of the human stomach mediates the competitive exclusion of the pathogen *H. pylori*.

The host-pathogen interaction studies in recalcitrant *Bifidobacterium breve* UCC2003 and *Bifidobacterium longum* 105-A have resulted in high transformation efficiency using PAM strategy and non-replicative plasmid-mediated homologous recombination [38, 45]. Carbohydrate uptake and metabolism, host cell adaptation, and immunomodulation have been extensively documented by generating single-crossover insertion mutants in *B. breve* UCC2003. The single-crossover insertion and double-crossover deletion in *B. longum* 105-A have established the molecular mechanisms of human milk oligosaccharides (HMOs), and co-evolution between *Bifidobacteria* and infants. The persistence of *Bifidobacteria* in the gut of breastfed infants is determined by several enzymes and transporters. The expansion of genome editing tools for *Bifidobacteria* facilitates understanding the interactions between the host and probiotics.

Bioengineer Tailored Probiotics by Genome Editing

The development of safe and robust probiotics for food and pharmaceutical applications includes modification in their genomes such as insertion, deletion, and activation using genome editing tools. The prevalence of antibiotic resistance genes such as *tetW* flanked by mobile genetic elements is high in *Bifidobacteria* [46]. The applications of GE tools can delete transferable antibiotic-resistance genes in potential probiotic candidates. *Bifidobacteria* have diverse prophage regions that can be activated to produce active phage leading to cell lysis and can cause contaminations in industrial fermentation. The cell lysis can be prevented by prophage curing, and deletion of related genes to enhance genetic stability of the strains. However, deletion of other mobile elements may also be possible.

The health beneficial properties conferred by *Bifidobacteria*, make them ideal candidates to be engineered for the delivery of therapeutic proteins either by cell surface display or extracellular secretion [47]. The use of an integrative expression system for both cell surface display and secretion of antibodies in human vaginal probiotics prevents HIV-1 infection. In addition, *Bifidobacteria* can also be engineered to improve gastrointestinal tract (GIT) stress tolerance and withstand food processing. A study integrated a catalase gene *katL* from the *L. plantarum* strain to the downstream of the *hup* gene in the chromosome of *B. longum* NCC2705. The catalase expression driven by the native *hup* promoter conferred increased viability under the H_2O_2 challenge in the recombinant strain [34].

Genome editing can be used for point mutation, gene replacement, and deletion to repair naturally inactivated genes. *Bifidobacterium animalis* DSM10140^T ability to produce EPS could be restored by introducing a point mutation at the

Balat_1410 gene, involved in EPS chain elongation [48]. The restoration of galactose fermentation in *Streptococcus thermophilus* strain CNRZ 302 can be achieved through the activation of the silent promoter by point mutation and single base pair insertion. In addition, the bacteriocin production in *L. gasseri* DSM 14869 might be restored by editing the nonsense mutation in the putative two-component system response regulator.

The use of genome editing in *Bifidobacteria* may target promoters to engineer target sequences through high expression by replacing the native promoter with inducible promoters. Besides, there is a rise in attention towards probiotics for use in cell factories to produce food and pharmaceutical products. Improving the production of cell factories through metabolic engineering by altering pathway distributions and regulating gene expression is particularly relevant in *Bifidobacteria*. The engineered strains with tailored functionalities and improved viability may affect symbiotic microbes in the human GIT. Therefore, the strains may be designed to exert beneficial functions under specific conditions within a particular niche.

MEDICAL APPLICATIONS FOR BIFIDOBACTERIAL GENE EXPRESSION SYSTEMS

Bifidobacteria represent the ideal host for *in situ* delivery of active substances due to their safety and ability to colonize the human gastrointestinal tract. The applications of *Bifidobacterium* in biomedical applications were initiated after successful murine mucosal immunization to *Salmonella* through flagellin expression from *B. animalis* [49]. The *Bifidobacteria* strains are significant in tumor suppression studies.

Genetically modified strains have emerged as potential biological agents with natural tumor specificity. The *Bifidobacterium* sp. to target human cancers has been achieved through the localization of species to the hypoxic regions of tumors following intravenous application. This potentially targets systemic metastases and primary tumors. Further, the expression of heterologous genes using plasmid transfection can localize bacteria to tumors and express therapeutic proteins [50].

The cytosine deaminase (CD) enzyme converts the non-toxic prodrug 5-flurocytosine (5-FC) to chemotherapeutic 5-flurouracil (5-FU), systemically administered to treat solid tumors. However, in conventional therapy, its clinical effectiveness is limited by high systemic toxicity to bone marrow. The "Bifidobacterial Selective Targeting" BEST strategy was developed for therapeutic purposes and exploits *B. longum* 105-A containing pBLES100-S-eCD, a plasmid-based on the shuttle vector pBLES100 with a cytosine deaminase gene (CD) under the control of the *hup* gene promoter. Recombinant *B. longum* can produce a CD in hypoxic mammary tumor tissues in rats, where it converts 5-FC into 5-FU *in vivo*. An improved version of pBLES100-S-eCD with 10-fold increased CD activity in *B. longum* has been developed.

B. longum subsp. *infantis* employing pGEX-1λT vector has been reported to express CD to inhibit melanoma in mice. The pGEX-1λT vector contains only the pBR322 *ori*. *Bifidobacterium* sp. has also been reported to have ColE1 plasmids and use of ampicillin as an antibiotic marker for bifidobacterial transformants since it is inactive in Gram-positive bacteria. *B. infantis*-mediated thymidine kinase plus ganciclovir (BI-TK/GCV) transformation was performed using ColE1 plasmid as a suicide gene therapy system for bladder tumors in rat models [51].

The strains of *B. adolescentis* and *B. longum* have also been used as delivery vectors to treat liver cancer in mice by transport of anti-angiogenic endostatin. For endostatin expression use of pMB1 replicon instead of pBV220 vector for *B. longum* expression of TNF- related apoptosis-inducing ligand (TRAIL) for antitumor effect on murine osteosarcoma [52]. Therefore the use of *Bifidobacterium* as a gene delivery vector has no adverse effects in any *in vivo* animal models.

The oral administration of *B. breve* UCC2003 harboring pLuxMC3 vector when fed to mice affected with subcutaneous tumours, observed the presence of cells by live whole-body imaging and reporter gene expression in tumors. The results show that *Bifidobacteria* translocate from the gastrointestinal tract (GIT) and replicate specifically in tumours [53]. Therefore, the translocated *B. breve* UCC2003 is non-pathogenic even in immunocompromised animals and the potential for gene delivery to the tumour environment compares favorably with the existing vectors. This strategy represents a safe and novel, non-invasive vector system, for delivering therapeutics.

MUTAGENESIS OF BIFIDOBACTERIAL GENES

The first successful knockout in *B. breve* UCC2003 of amylopullulanase-encoding *apuB* gene by single crossover recombination employed two different strategies. The first is based on the lactococcal temperature-sensitive plasmid pVE6007 for disruption of apuB gene which is successfully achieved however, it is tedious and unreliable. The second approach is insertional mutagenesis of the *apuB* gene and the *galA,* endogalactanase gene of *B. breve* UCC2003, demonstrating high transformation efficiency due to plasmid methylation. The approach can be applied even for the homologous DNA of less than 500bp and allows for site-specific homologous recombination [54]. A gene deletion method by two homologous recombination events reportedly deleted *tet*(W), tetracycline resistance gene from *B. longum* subsp. *longum* NCC2705 and *B. animalis* subsp. *lactis* NCC2818 (Bb12) [55 - 57].

CONCLUSION

The scientific evidence on probiotics are increasing drastically and is gaining public attention. The potential to present bacteria in the intestine *via* a bacterial delivery and expression system needs to be accelerated. Genome engineering is the key to developing novel and reliable engineered bacteria that help understand the role of probiotics in human health. The engineered bacteria can be reprogrammed to enhance effector molecule production and for mucosal delivery of therapeutics. The maintenance of cellular and biological processes involves multiple enzymatic reactions, defects in these enzymes may result in essential metabolite production. The engineering of probiotics harboring specific enzymes relieves metabolic disorders in the organism.

The increasing availability of genetic tools and genome sequences for *Bifidobacteria* provides the opportunity for improved analysis of the physiology, genetics, and metabolism of the gut bacterium and its role in gastrointestinal health. The improvement of probiotics functionality has observed metabolic engineering approaches in the usage of food-grade selection markers, implementation of bifidobacterial genes, and chromosomal integration. The independent strains developed which are non-pathogenic to the environment have paved the way for clinical applications. Synthetic biology approaches along with genome editing tools will boost the development of next-generation microbial therapeutics. Further, the increasing number of novel applications and considerable advantages of *Bifidobacteria* will enhance the therapeutic and industrial importance.

ACKNOWLEDGEMENTS

The Authors would like to thank the Director CSIR- Central Food Technological Research Institute for providing the facilities to carry out the research. AS would like to thank CSIR for awarding Nehru Post-Doctoral fellowship.

REFERENCES

[1]   Clemente JC, Ursell LK, Parfrey LW, Knight R. The impact of the gut microbiota on human health: An integrative view. Cell 2012; 148(6): 1258-70.
[http://dx.doi.org/10.1016/j.cell.2012.01.035] [PMID: 22424233]

[2]   David LA, Maurice CF, Carmody RN, *et al.* Diet rapidly and reproducibly alters the human gut microbiome. Nature 2014; 505(7484): 559-63.
[http://dx.doi.org/10.1038/nature12820] [PMID: 24336217]

[3]   Qin J, Li Y, Cai Z, *et al.* A metagenome-wide association study of gut microbiota in type 2 diabetes. Nature 2012; 490(7418): 55-60.

[http://dx.doi.org/10.1038/nature11450] [PMID: 23023125]

[4]   Benson AK, Kelly SA, Legge R, *et al.* Individuality in gut microbiota composition is a complex polygenic trait shaped by multiple environmental and host genetic factors. Proc Natl Acad Sci 2010; 107(44): 18933-8.
[http://dx.doi.org/10.1073/pnas.1007028107] [PMID: 20937875]

[5]   Ventura M, Turroni F, Van Sinderen D. Probiogenomics as a tool to obtain genetic insights into adaptation of probiotic bacteria to the human gut. Bioeng Bugs 2012; 3(2): 73-9.
[PMID: 22095053]

[6]   Lugli GA, Milani C, Turroni F, *et al.* Investigation of the evolutionary development of the genus *Bifidobacterium* by comparative genomics. Appl Environ Microbiol 2014; 80(20): 6383-94.
[http://dx.doi.org/10.1128/AEM.02004-14] [PMID: 25107967]

[7]   Bazanella M, Maier TV, Clavel T, *et al.* Randomized controlled trial on the impact of early-life intervention with *Bifidobacteria* on the healthy infant fecal microbiota and metabolome. Am J Clin Nutr 2017; 106(5): 1274-86.
[http://dx.doi.org/10.3945/ajcn.117.157529] [PMID: 28877893]

[8]   Milani C, Lugli GA, Duranti S, *et al.* Genomic encyclopedia of type strains of the genus *Bifidobacterium.* Appl Environ Microbiol 2014; 80(20): 6290-302.
[http://dx.doi.org/10.1128/AEM.02308-14] [PMID: 25085493]

[9]   Pokusaeva K, Fitzgerald GF, Van Sinderen D. Carbohydrate metabolism in *Bifidobacteria.* Genes Nutr 2011; 6(3): 285-306.
[http://dx.doi.org/10.1007/s12263-010-0206-6] [PMID: 21484167]

[10]   Sánchez B, Urdaci MC, Margolles A. Extracellular proteins secreted by probiotic bacteria as mediators of effects that promote mucosa–bacteria interactions. Microbiology 2010; 156(11): 3232-42.
[http://dx.doi.org/10.1099/mic.0.044057-0] [PMID: 20864471]

[11]   Fukuda S, Toh H, Hase K, *et al. Bifidobacteria* can protect from enteropathogenic infection through production of acetate. Nature 2011; 469(7331): 543-7.
[http://dx.doi.org/10.1038/nature09646] [PMID: 21270894]

[12]   Charnchai P, Jantama SS, Prasitpuriprecha C, Kanchanatawee S, Jantama K. Effects of the Food manufacturing chain on the viability and functionality of *Bifidobacterium animalis* through simulated gastrointestinal conditions. PLoS One 2016; 11(6): e0157958.
[http://dx.doi.org/10.1371/journal.pone.0157958] [PMID: 27333286]

[13]   Picard C, Fioramonti J, Francois A, Robinson T, Neant F, Matuchansky C. Review article: *Bifidobacteria* as probiotic agents - physiological effects and clinical benefits. Aliment Pharmacol Ther 2005; 22(6): 495-512.
[http://dx.doi.org/10.1111/j.1365-2036.2005.02615.x] [PMID: 16167966]

[14]   Heinken A, Ravcheev DA, Baldini F, Heirendt L, Fleming RMT, Thiele I. Systematic assessment of secondary bile acid metabolism in gut microbes reveals distinct metabolic capabilities in inflammatory bowel disease. Microbiome 2019; 7(1): 75.
[http://dx.doi.org/10.1186/s40168-019-0689-3] [PMID: 31092280]

[15]   Thiele I, Heinken A, Fleming RMT. A systems biology approach to studying the role of microbes in human health. Curr Opin Biotechnol 2013; 24(1): 4-12.
[http://dx.doi.org/10.1016/j.copbio.2012.10.001] [PMID: 23102866]

[16]   Shoaie S, Ghaffari P, Kovatcheva-Datchary P, *et al.* Quantifying diet-induced metabolic changes of the human gut microbiome. Cell Metab 2015; 22(2): 320-31.
[http://dx.doi.org/10.1016/j.cmet.2015.07.001] [PMID: 26244934]

[17]   Hamilton-Miller JMT. The role of probiotics in the treatment and prevention of *Helicobacter pylori* infection. Int J Antimicrob Agents 2003; 22(4): 360-6.
[http://dx.doi.org/10.1016/S0924-8579(03)00153-5] [PMID: 14522098]

[18]   Sgouras D, Maragkoudakis P, Petraki K, *et al*. *In vitro* and *in vivo* inhibition of *Helicobacter pylori* by *Lactobacillus casei* strain Shirota. Appl Environ Microbiol 2004; 70(1): 518-26.
[http://dx.doi.org/10.1128/AEM.70.1.518-526.2004] [PMID: 14711683]

[19]   Ventura M, O'Flaherty S, Claesson MJ, *et al*. Genome-scale analyses of health-promoting bacteria: Probiogenomics. Nat Rev Microbiol 2009; 7(1): 61-71.
[http://dx.doi.org/10.1038/nrmicro2047] [PMID: 19029955]

[20]   Sela DA, Chapman J, Adeuya A, *et al*. The genome sequence of *Bifidobacterium longum* subsp. *infantis* reveals adaptations for milk utilization within the infant microbiome. Proc Natl Acad Sci 2008; 105(48): 18964-9.
[http://dx.doi.org/10.1073/pnas.0809584105] [PMID: 19033196]

[21]   Pompei A, Cordisco L, Amaretti A, Zanoni S, Matteuzzi D, Rossi M. Folate production by *Bifidobacteria* as a potential probiotic property. Appl Environ Microbiol 2007; 73(1): 179-85.
[http://dx.doi.org/10.1128/AEM.01763-06] [PMID: 17071792]

[22]   Turroni F, Ribbera A, Foroni E, Van Sinderen D, Ventura M. Human gut microbiota and *Bifidobacteria*: From composition to functionality. Antonie van Leeuwenhoek 2008; 94(1): 35-50.
[http://dx.doi.org/10.1007/s10482-008-9232-4] [PMID: 18338233]

[23]   Zomer A, Fernandez M, Kearney B, Fitzgerald GF, Ventura M, Van Sinderen D. An interactive regulatory network controls stress response in *Bifidobacterium breve* UCC2003. J Bacteriol 2009; 191(22): 7039-49.
[http://dx.doi.org/10.1128/JB.00897-09] [PMID: 19734308]

[24]   Cronin M, Ventura M, Fitzgerald GF, Van Sinderen D. Progress in genomics, metabolism and biotechnology of *Bifidobacteria*. Int J Food Microbiol 2011; 149(1): 4-18.
[http://dx.doi.org/10.1016/j.ijfoodmicro.2011.01.019] [PMID: 21320731]

[25]   Pokusaeva K, O'Connell-Motherway M, Zomer A, Fitzgerald GF, Van Sinderen D. Characterization of two novel α-glucosidases from *Bifidobacterium breve* UCC2003. Appl Environ Microbiol 2009; 75(4): 1135-43.
[http://dx.doi.org/10.1128/AEM.02391-08] [PMID: 19114534]

[26]   Newburg DS. Neonatal protection by an innate immune system of human milk consisting of oligosaccharides and glycans1. J Anim Sci 2009; 87(13): 26-34.
[http://dx.doi.org/10.2527/jas.2008-1347] [PMID: 19028867]

[27]   Le Leu RK, Hu Y, Brown IL, Woodman RJ, Young GP. Synbiotic intervention of *Bifidobacterium lactis* and resistant starch protects against colorectal cancer development in rats. Carcinogenesis 2010; 31(2): 246-51.
[http://dx.doi.org/10.1093/carcin/bgp197] [PMID: 19696163]

[28]   Zihler A, Gagnon M, Chassard C, Lacroix C. Protective effect of probiotics on Salmonella infectivity assessed with combined *in vitro* gut fermentation-cellular models. BMC Microbiol 2011; 11(1): 264.
[http://dx.doi.org/10.1186/1471-2180-11-264] [PMID: 22171685]

[29]   Tanner SA, Chassard C, Zihler Berner A, Lacroix C. Synergistic effects of *Bifidobacterium thermophilum* RBL67 and selected prebiotics on inhibition of Salmonella colonization in the swine proximal colon PolyFermS model. Gut Pathog 2014; 6(1): 44.
[http://dx.doi.org/10.1186/s13099-014-0044-y] [PMID: 25364390]

[30]   Wolvers D, Antoine JM, Myllyluoma E, Schrezenmeir J, Szajewska H, Rijkers GT. Guidance for substantiating the evidence for beneficial effects of probiotics: Prevention and management of infections by probiotics. J Nutr 2010; 140(3): 698S-712S.
[http://dx.doi.org/10.3945/jn.109.113753] [PMID: 20107143]

[31]   Yun B, Song M, Park DJ, Oh S. Beneficial effect of *Bifidobacterium longum* ATCC 15707 on survival rate of *Clostridium difficile* infection in mice. Han-gug Chugsan Sigpum Hag-hoeji 2017; 37(3): 368-75.

[http://dx.doi.org/10.5851/kosfa.2017.37.3.368] [PMID: 28747822]

[32] Hirayama Y, Sakanaka M, Fukuma H, *et al.* Development of a double-crossover markerless gene deletion system in *Bifidobacterium longum*: Functional analysis of the α-galactosidase gene for raffinose assimilation. Appl Environ Microbiol 2012; 78(14): 4984-94.
[http://dx.doi.org/10.1128/AEM.00588-12] [PMID: 22582061]

[33] Xin Y, Guo T, Mu Y, Kong J. Identification and functional analysis of potential prophage-derived recombinases for genome editing in *Lactobacillus casei*. FEMS Microbiol Lett 2017; 364(24).
[http://dx.doi.org/10.1093/femsle/fnx243] [PMID: 29145601]

[34] Zuo F, Zeng Z, Hammarström L, Marcotte H. Inducible plasmid self-destruction (ipsd) assisted genome engineering in *Lactobacilli* and *Bifidobacteria*. ACS Synth Biol 2019; 8(8): 1723-9.
[http://dx.doi.org/10.1021/acssynbio.9b00114] [PMID: 31277549]

[35] Sakaguchi K, Funaoka N, Tani S, *et al.* The pyrE gene as a bidirectional selection marker in *Bifidobacterium longum* 105-A. Biosci Microbiota Food Health 2013; 32(2): 59-68.
[http://dx.doi.org/10.12938/bmfh.32.59] [PMID: 24936363]

[36] Oh JH, Van Pijkeren JP. CRISPR–Cas9-assisted recombineering in *Lactobacillus reuteri*. Nucleic Acids Res 2014; 42(17): e131-1.
[http://dx.doi.org/10.1093/nar/gku623] [PMID: 25074379]

[37] Song X, Huang H, Xiong Z, Ai L, Yang S. CRISPR-Cas9 D10A nickase-assisted genome editing in *Lactobacillus casei*. Appl Environ Microbiol 2017; 83(22): e01259-17.
[http://dx.doi.org/10.1128/AEM.01259-17] [PMID: 28864652]

[38] Genetic manipulation and gene modification technologies in *Bifidobacteria*. The *Bifidobacteria* and related organisms. Academic Press 2022; 243-59.

[39] Pan M, Nethery MA, Hidalgo-Cantabrana C, Barrangou R. Comprehensive mining and characterization of CRISPR-cas systems in *bifidobacterium*. Microorganisms 2020; 8(5): 720.
[http://dx.doi.org/10.3390/microorganisms8050720] [PMID: 32408568]

[40] Vento JM, Crook N, Beisel CL. Barriers to genome editing with CRISPR in bacteria. J Ind Microbiol Biotechnol 2019; 46(9-10): 1327-41.
[http://dx.doi.org/10.1007/s10295-019-02195-1] [PMID: 31165970]

[41] Wang C, Cui Y, Qu X. Optimization of electrotransformation (ETF) conditions in lactic acid bacteria (LAB). J Microbiol Methods 2020; 174: 105944.
[http://dx.doi.org/10.1016/j.mimet.2020.105944] [PMID: 32417130]

[42] Grimm V, Gleinser M, Neu C, Zhurina D, Riedel CU. Expression of fluorescent proteins in *Bifidobacteria* for analysis of host-microbe interactions. Appl Environ Microbiol 2014; 80(9): 2842-50.
[http://dx.doi.org/10.1128/AEM.04261-13] [PMID: 24584243]

[43] Zeng Z, Zuo F, Marcotte H. Putative adhesion factors in vaginal *Lactobacillus gasseri* DSM 14869: Functional characterization. Appl Environ Microbiol 2019; 85(19): e00800-19.
[http://dx.doi.org/10.1128/AEM.00800-19] [PMID: 31420338]

[44] Goh YJ, Azcárate-Peril MA, O'Flaherty S, *et al.* Development and application of a *upp*-based counterselective gene replacement system for the study of the S-layer protein SlpX of *Lactobacillus acidophilus* NCFM. Appl Environ Microbiol 2009; 75(10): 3093-105.
[http://dx.doi.org/10.1128/AEM.02502-08] [PMID: 19304841]

[45] Kanesaki Y, Masutani H, Sakanaka M, *et al.* Complete genome sequence of *Bifidobacterium longum* 105-A, a strain with high transformation efficiency. Genome Announc 2014; 2(6): e01311-14.
[http://dx.doi.org/10.1128/genomeA.01311-14] [PMID: 25523770]

[46] Rozman V, Mohar Lorbeg P, Accetto T, Bogovič Matijašić B. Characterization of antimicrobial resistance in *Lactobacilli* and *Bifidobacteria* used as probiotics or starter cultures based on integration of phenotypic and *in silico* data. Int J Food Microbiol 2020; 314: 108388.

[http://dx.doi.org/10.1016/j.ijfoodmicro.2019.108388] [PMID: 31707173]

[47]   Del Rio B, Redruello B, Fernandez M, Martin MC, Ladero V, Alvarez MA. Lactic acid bacteria as a live delivery system for the *in situ* production of nanobodies in the human gastrointestinal tract. Front Microbiol 2019; 9: 3179.
[http://dx.doi.org/10.3389/fmicb.2018.03179]

[48]   Hidalgo-Cantabrana C, Sánchez B, Álvarez-Martín P, *et al.* A single mutation in the gene responsible for the mucoid phenotype of *Bifidobacterium animalis* subsp. lactis confers surface and functional characteristics. Appl Environ Microbiol 2015; 81(23): 7960-8.
[http://dx.doi.org/10.1128/AEM.02095-15] [PMID: 26362981]

[49]   Takata T, Shirakawa T, Kawasaki Y, *et al.* Genetically engineered *Bifidobacterium animalis* expressing the *Salmonella flagellin* gene for the mucosal immunization in a mouse model. J Gene Med 2006; 8(11): 1341-6.
[http://dx.doi.org/10.1002/jgm.963] [PMID: 16958059]

[50]   Baban CK, Cronin M, O'Hanlon D, O'Sullivan GC, Tangney M. Bacteria as vectors for gene therapy of cancer. Bioeng Bugs 2010; 1(6): 385-94.
[http://dx.doi.org/10.4161/bbug.1.6.13146] [PMID: 21468205]

[51]   Tang W, He Y, Zhou S, Ma Y, Liu G. A novel *Bifidobacterium infantis*-mediated TK/GCV suicide gene therapy system exhibits antitumor activity in a rat model of bladder cancer. J Exper Clin Cancer Res 2009; 28(1).

[52]   Xu Y-F, Zhu L-P, Hu B, *et al.* A new expression plasmid in *Bifidobacterium longum* as a delivery system of endostatin for cancer gene therapy. Cancer Gene Ther 2007; 14(2): 151-7.
[http://dx.doi.org/10.1038/sj.cgt.7701003] [PMID: 17068487]

[53]   Cronin M, Morrissey D, Rajendran S, *et al.* Orally administered *Bifidobacteria* as vehicles for delivery of agents to systemic tumors. Mol Ther 2010; 18(7): 1397-407.
[http://dx.doi.org/10.1038/mt.2010.59] [PMID: 20389288]

[54]   O'Connell Motherway M, O'Driscoll J, Fitzgerald GF, Van Sinderen D. Overcoming the restriction barrier to plasmid transformation and targeted mutagenesis in *Bifidobacterium breve* UCC2003. Microb Biotechnol 2009; 2(3): 321-32.
[http://dx.doi.org/10.1111/j.1751-7915.2008.00071.x] [PMID: 21261927]

[55]   Arigoni F. Genetic remodeling in *Bifidobacterium*. W.O. Patent 2008/019886A1, 2022.

[56]   Sundararaman A, Halami PM. Genome editing of probiotic bacteria: Present status and future prospects. Biologia 2022; 77(7): 1831-41.
[http://dx.doi.org/10.1007/s11756-022-01049-z]

[57]   Sundararaman A, Bansal K, Sidhic J, Patil P, Halami PM. Genome of *Bifidobacterium longum* NCIM 5672 provides insights into its acid-tolerance mechanism and probiotic properties. Arch Microbiol 2021; 203(10): 6109-18.
[http://dx.doi.org/10.1007/s00203-021-02573-3] [PMID: 34553262]

CHAPTER 6

Lactic Acid Bacteria as Starter Cultures in Food: Genome Characterization and Comparative Genomics

Md Minhajul Abedin[1], Srichandan Padhi[2], Rounak Chourasia[2], Loreni Chiring Phukon[1], Puja Sarkar[2], Sudhir P. Singh[3] and Amit Kumar Rai[1,*]

[1] *National Agri-Food Biotechnology Institute (DBT-NABI), S.A.S. Nagar, Mohali, Punjab, India*

[2] *Institute of Bioresources and Sustainable Development, Regional Centre, Tadong, Sikkim, India*

[3] *Center of Innovative and Applied Bioprocessing (DBT-CIAB), S.A.S. Nagar, Mohali, India*

Abstract: Fermented food products are consumed by about 30% of the world's population due to their high nutritional value and health properties. The use of LAB in the fermentation process has resulted in a variety of fermented food products derived from both plant and animal sources. LAB have been used as starter cultures for food fermentation both traditionally and industrially, having certain specific characteristics such as rapid growth, product yield, higher biomass and also unique organoleptic properties, and are employed in food fermentation. The advancement of high-throughput genome sequencing methods has resulted in a tremendous improvement in our understanding of LAB physiology and has become more essential in the field of food microbiology. The complete genome sequence of *Lactococcus lactis* in 2001 resulted in a better understanding of metabolic properties and industrial applications of LAB. Genes associated with β-galactosidase, antimicrobial agents, bile salt hydrolase, exopolysaccharide, and GABA producing LAB have received a lot of attention in recent years. Genome editing techniques are required for the development of strains for novel applications and products. They can also play an important part as a research method for acquiring mechanistic insights and identifying new properties. The genome editing of lactic acid bacterial strains has a lot of potential applications for developing functional foods with a favourable influence on the food industries.

Keywords: Functional food, Genes, Genome editing, Lactic acid bacteria, LAB genome, Metagenomics, Sequencing-approaches, Starter culture.

---

[*] **Corresponding author Amit Kumar Rai:** National Agri-Food Biotechnology Institute (DBT-NABI), S.A.S. Nagar, Mohali, Punjab, India; E-mail: amitraikvs@gmail.com

Prakash M. Halami & Aravind Sundararaman (Eds.)

INTRODUCTION

Fermentation is one of the oldest and cheapest methods of food processing. To date, it is estimated that globally more than 5,000 varieties of different fermented beverages and foods are being consumed. Traditional fermented food products such as cheese, soya sauce, meats, vegetables and wine production have been documented since ancient civilization [1]. The process of fermentation improves the nutritional content and shelf-life, also increasing the functionality and therapeutic effects of the end product. The process further reduces anti-nutritional properties such as tannins and phytic acid, enhancing the bioactive components which exert beneficial health effects along with increasing the flavour, texture and aroma [2, 3]. Bioactive molecules produced during the fermentation of food have gained attention in the research due to their specific health promoting properties, having antioxidant, anti-cholesterol, anti-inflammatory, immunomodulatory, ACE-inhibitory and anti-viral properties [4 - 7].

Lactic acid bacteria (LAB) are Gram-positive, facultative anaerobic, non-sporulating and catalase negative bacteria, involved in the production of fermented foods such as meat, vegetables, cereals and dairy products [8]. Common bacterial genera of LAB include *Lactobacillus, Lactococcus, Enterococcus, Leuconostoc, Bifidobacterium, Pediococcus, Propionibacterium, Streptococcus, Weissella, Tertragenococcus* and *Vagococcus*. According to the new classification of bacterial taxonomy, LAB belongs to the phylum *Firmicutes*, class *Bacilli*, order *Lactobacillales*, and refers to five families, which include *Aerococcaceae, Carnobacteriaceae, Enterococcaceae, Lactobacillaceae* (includes the family *Leuconostocaceae*), and *Streptococcaceae* [9]. LAB are generally considered safe (GRAS) and most of them have probiotic effect [8, 10]. LAB have been used as starters, which is a single and/or mixed culture employed to enhance the rate of the food fermentation process to provide particular characteristics in a controlled manner. LAB are quite frequently used along with fungi and yeast during the process of fermentation, which helps in the release of novel bioactive compounds [11]. Production of lactic acid resulting in acidification of the food is the primary function of starter culture. The use of starter culture has resulted in an increase of food safety by inhibiting the growth of undesirable microorganisms which leads to spoilage of food products and also pathogenic microbes. Since the early 20th century, numerous commercial starter cultures have been employed which possess different metabolic properties, including proteolytic, acidification and antagonistic properties [8].

In the field of food microbiology, microbial genome sequencing has been playing an important role because of its improvement in the speed and quality of sequencing data. Discovery of complete genome sequencing of *Lactococcus lactis* subsp. *lactis* IL1403 in 2001, have provided new insight and application for LAB [12]. More than 100 complete genomes of LAB have been sequenced and are available in the public domain [13] (Fig. **1**) (Table **1**). Developments of genomics and functional genomics, along with the high throughput technologies in the last decade, have resulted in a greater understanding of metabolic characteristics and industrial use of LAB, thus, making LAB as the most promising microorganisms for food and other industries by large scale production of food ingredients and other products such as polyols, lactic acid and vitamins.

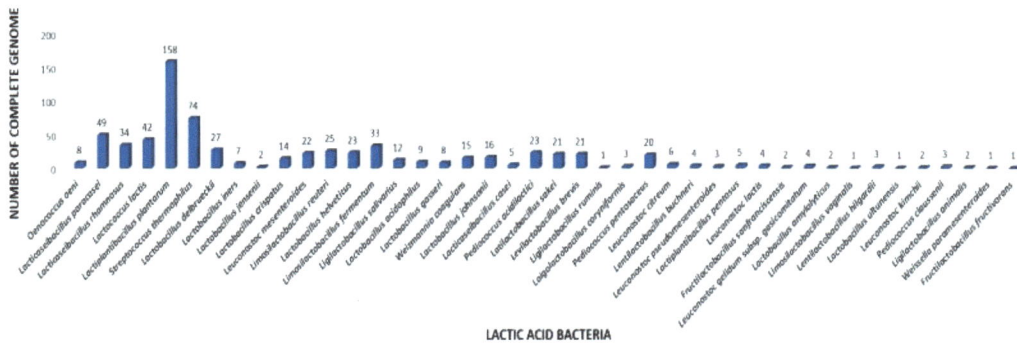

Fig. (1). Number of a complete genome sequence of various LAB available in the public database.

Table 1. Selected LAB species and their genome information retrieved from the available literature.

LAB Species	Genome Size (Mb)	Genes	GC%	References
Lactobacillus acidophilus La-14	1.99	1957	34.7	[85]
Lb. acidophilus NCFM	1.99	1938	34.7	[86]
Levilactobacillus brevis ATCC 367	2.29	2314	46.2	[87]
Lacticaseibacillus casei ATCC 334	2.76	2906	46.7	[87]
Lactobacillus delbrueckii subsp. *bulgaricus* 2038	1.87	1792	49.7	[88]
Lb. delbrueckii subsp. *bulgaricus* ND02	2.13	2183	49.6	[89]
Limosilactobacillus fermentum IFO 3956	2.00	1912	51.0	[90]
Lactobacillus gasseri ATCC 33323	1.95	1898	35	[91]
Lactobacillus helveticus DPC 4571	1.98	1938	37.0	[92]
Lactobacillus johnsonii NCC 533	1.99	1918	34.6	[93]
Lactiplantibacillus plantarum WCFS1	3.31	3135	44.5	[94]
Lacticaseibacillus rhamnosus GG	3.01	2905	46.7	[95]
Latilactobacillus sakei subsp. *sakei* 23k	1.90	1963	41.2	[96]
Ligilactobacillus salivarius UCC118	1.83	1864	32.9	[97]

(Table 1) cont.....

LAB Species	Genome Size (Mb)	Genes	GC%	References
Lc. lactis IL1403	2.37	2425	35.4	[12]
Lc. lactis subsp. *cremoris* MG1363	2.53	2597	35.8	[74]
Leuconostoc kimchii IMSNU 11154	2.10	2209	37.9	[98]
Leuconostoc mesenteroides subsp. mesenteroides ATCC 8293	1.94	2073	37.0	[87]
Streptococcus thermophilus LMG18311	1.80	1973	39.1	[99]
S. thermophilus MN-ZLW-002	1.85	2046	39.1	[100]

In recent years, various beneficial properties of LAB, such as β-galactosidase, antimicrobial agents, bile salt hydrolase, exopolysaccharide, and GABA, have gained a lot of attention. LAB producing β-galactosidases are employed in food technology to hydrolyse lactose in milk because of the prevalence of lactose intolerance in the human population and the relevance of milk in the human diet [14, 15]. The use of LAB has been employed in several fermented foods products. This is so because of their antimicrobial activities, which inhibit the growth of harmful microbes and pathogens, resulting in the prevention of spoilage [16]. Several GABA producing LAB have been isolated from fermented food products and have been employed to produce foods enriched with GABA. GABA functions as a key inhibitory neurotransmitter in the mammalian central nervous system [17]. LAB are a significant part of the Firmicutes bacteria found in the human gut, and bile salt hydrolase activity has been shown in numerous lactic acid bacterial species. Bile salt hydrolase catalyses the hydrolysis of glycine and/or taurine-conjugated bile salts, resulting in the release of free bile acids and amino acids [18]. Several LAB have been extensively investigated for their capacity to generate exopolysaccharides which are repeating units of mono/oligosaccharides with varying chemical compositions and characteristics. EPS promotes apoptosis, causing cell cycle arrest, and also has anti-mutagenic, anti-oxidative, anti-angiogenesis, and anti-inflammatory properties [8].

One of the fundamental tools for LAB engineering is genome editing, which involves inserting or removing target genes from the host. Significant progress has been made in LAB engineering since the advent of synthetic biology. LAB engineering through synthetic biology is currently in the attention due to growing interest in their use as probiotics. The benefits of native probiotics, combined with the advancement of synthetic biology, have aided in the development of LAB into the next generation of living therapeutics, capable of performing specific actions in the human body, such as preventing and fighting infections, removing tumours, and treating metabolic disorders. Synthetic biological genetic tools have recently been proven to play a critical role in probiotic engineering, paving the way for the successful engineering of fully functional therapeutics. Genetically engineered LAB have been employed in various industries, such as dairy and other food fermentation industries. Several attempts have been made to develop effective vector systems for LAB metabolic engineering, such as *Lc. lactis* and *S. thermophilus* have used genetically engineered LAB to produce B-vitamins, diacetyl, acetaldehyde, and folate [10].

This chapter briefly describes the LAB used in food, starter culture and its selection, characterization and different approaches used for genome sequencing, practical insights into LAB genomes highlighting the key genes involved and finally applications of genome editing in LAB.

Popular LAB used in Food

From ancient times, the fermentation of food has become an essential method for human beings to enrich the dietary intake of foods and also for preservation. It has been found that LAB was used to ferment milk, mostly during the Indus Valley, Babylonian and Egyptian civilization. Traditional fermented foods have been consumed for centuries, possessing beneficial effects on communities and having historical evidence [1]. The use of LAB in food fermentation plays an important role as they result in diverse textures, aromas and tastes. Often productions of traditional fermented foods are considered unsafe and unsanitary, though it occurs at some stage but is overstated. It has been reported that due to the low moisture, high acidity and salt concentration fermentation process can inhibit the growth of spoilage bacteria [19]. Traditional fermented foods are produced by the process of natural fermentation that does not require sterilization. LAB are widely used in vegetable fermentation for preservation and its history can be found in many Asian and European countries [19].

In the present day, around 20-30% of the population around the globe consumes fermented food products due to their rich nutrition and unique flavour [1] (Table **2**). Common LAB fermented food products include cheese, yogurt, sourdough, sauerkraut, olives, sausages, kimchi, pickled vegetables, steamed bun, fermented rice and soybean [8]. *Aerococcus*, *Carnobacterium*, *Enterococcus*, *Lactobacillus*, *Lactococcus*, *Leuconostoc*, *Oenococcus*, *Pediococcus*, *Sporolactobacillus*, *Streptococcus*, *Tetragenococcus*, *Vagococcus*, and *Weissella* are among the LAB found in fermented food products.

Fermented Plant Based Food Product

Fermentation of plant-based products, such as vegetables, as a method of preservation and consumption has a long history, which dates back to the early human civilization. Consumption of sauerkraut during the Roman Empire or the traditional sauerkrauts from Asia, such as *Kimchi* of Korea, and also consumption of sourdough by the Mediterranean civilizations are the prime examples of traditional fermented plant based food products using LAB [20]. Demand for the consumption of fermented vegetables has received great attention in developed countries especially due to its health benefits. Vegetables are the sources of various nutrients, such as minerals, vitamins, prebiotic fibres and other phytochemicals. *Enterococcus, Lactobacillus, Leuconostoc, Pediococcus*, and *Weissella* are the common LAB which are involved in the lactic acid fermentation of vegetables [21].

Table 2. Details of LAB strains whose genomes are available in the public databases.

LAB Species	Strain	Source Foods	Geographic Location	NCBI BioSample ID
Oenococcus oeni	SD-2a	Wine	China	11269073
O. oeni	19	Patagonian red wine	Argentina	8545569
Lacticaseibacillus paracasei	NJ	Yogurt	China	12319598
La. paracasei	TCS	Milk fan	China	11177072
La. paracasei	Lpc10	Patagonian merlot wine	Argentina	9283999
La. paracasei	HD1.7	Chinese sauerkraut	China	8195363
La. paracasei	TK1501	Congee	China	5921089
La. paracasei	EG9	Cheese	Japan	9239981
La. paracasei	FAM18149	Cheese	Switzerland	5781319
La. paracasei subsp. paracasei	TMW 1.1434	Starter culture	Germany	5356834
La. paracasei subsp. tolerans	MGB0625	Kimchi	South Korea	16668564
La. paracasei subsp. tolerans	ZY-1	Tibetan kefir	China	16861159
La. paracasei subsp. tolerans	S-NB	Fermented milk	China	17359333
La. paracasei subsp. tolerans	2A	Yogurt	USA	18869149
La. rhamnosus	KF7	Kefir	China	17214539
La. rhamnosus	X253	Fermented milk	China	18868053
La. rhamnosus	CE1	Cheese	USA	18687446
La. rhamnosus	AS	Cheese	USA	18869154
La. rhamnosus	B6	Yogurt	Bulgaria	17141195
Lc. lactis	N8	Cheese product	Finland	15500618
Lc. lactis	CBA3619	Kimchi	South Korea	11843663
Lc. lactis subsp. lactis	223	Milk starter culture	Ireland	9847869

(Table 2) cont.....

LAB Species	Strain	Source Foods	Geographic Location	NCBI BioSample ID
Lc. lactis subsp. lactis	UC11	Fermented Meat Product	Italy	4956294
Lc. lactis subsp. lactis	KLDS 4.0325	Home-made koumiss	China	2603468
Lc. lactis subsp. lactis bv. diacetylactis	S50	Butter starter	Serbia	10167144
Lc. lactis subsp. lactis biovar diacetylactis	FM03	Cheese	Denmark	6061939
Lp. plantarum	CNEI-KCA4	Fermented Okpei-Onitsha	Nigeria	14941080
Lp. plantarum	LP2	Pickles	China	6756312
Lp. plantarum	PC518	Sichuan pickle	China	16812448
Lp. plantarum	SK151	Kimchi	South Korea	9373895
Lp. plantarum	NCU116	Chinese pickle	China	5162259
Lp. plantarum	BK-021	Fermented onion	South Korea	11650980
Lp. plantarum	PC520	Chinese fermented food pickle	China	7760023
Lp. plantarum	K259	Kimchi	South Korea	8274873
Lp. plantarum	10CH	Cheese	UK	6311425
Lp. plantarum	ZDY2013	Traditional Chinese fermented soybeans	China	16814808
S. thermophilus	24739	Cheese starter culture	Switzerland	15923548
S. thermophilus	KLDS 3.1003	Traditional yogurt	China	5465222
S. thermophilus	ND07	Fermented yak milk	China	5357478
S. thermophilus	CS9	Fermented milk	China	9536719
S. thermophilus	MN-BM-A02	Dairy fan	China	3325854
Lb. delbrueckii	TS1-06	Fermented milk product "Chaka"	Tajikistan	13382266
Lb. delbrueckii subsp. bulgaricus	LJJ	Traditional fermented yogurt	China	14132428
Lb. delbrueckii subsp. bulgaricus	ND04	Fermented camel milk	China	5357477
Lb. delbrueckii subsp. bulgaricus	L99	Bulgarian homemade fermented milk	Bulgaria	5751687
Lb. delbrueckii subsp. delbrueckii	TUA4408L	Sunki	Japan	6812572

(Table 2) cont.....

LAB Species	Strain	Source Foods	Geographic Location	NCBI BioSample ID
Lb. delbrueckii subsp. lactis	MAG_rmk202_ldel	Cheese starter culture	Switzerland	13284946
Lb. delbrueckii	NWC_1_2	Natural whey culture from Gruyere cheese	Switzerland	9041770
Leu. mesenteroides	SRCM102733	Gochujang	South Korea	8707608
Leu. mesenteroides	WiKim32	Baechu-kimchi	South Korea	11026499
Leu. mesenteroides	SRCM102735	Chonggugjang	South Korea	8707610
Leu. mesenteroides	CBA3628	Kimchi	South Korea	11843664
Leu. mesenteroides subsp. dextranicum	DSM 20484	Cheese	Netherlands	3105774
Leu. mesenteroides subsp. jonggajibkimchii	DRC1506	Kimchi	South Korea	4198570
Leu. mesenteroides subsp. mesenteroides	BD1710	Kefir	China	4530070
Leu. mesenteroides subsp. mesenteroides	BD3749	Fermeted vegetable	China	4508169
Leu. mesenteroides subsp. mesenteroides	DRC0211	Kimchi	South Korea	4198273
Leu. mesenteroides subsp. mesenteroides	FM06	Samso cheese	Denmark	6061940
Limosilactobacillus reuteri	WHH1689	Highland barley wine	China	8382659
Lb. helveticus	FAM8105	Tilsit cheese	Switzerland	4892754
Lb. helveticus	CAUH18	Koumiss	China	3982146
Lb. helveticus	LZ-R-5	Yogurt	China	12670384
Lb. helveticus	LH99	Homemade fermented milk	Bulgaria	5751674
Lb. helveticus	NWC_2_4	Natural whey culture from Gruyere cheese	Switzerland	9476689
Lm. fermentum	LMT2-75	Kimchi	South Korea	10417155
Lm. fermentum	USM 8633	Fermented meat sausage	Malaysia	4531192
Lm. fermentum	GR1007	Fermented meat sausages	Taiwan	17976236
Lm. fermentum	B44	Raw milk	China	17214540
Lm. fermentum	ERS1883068	Sourdough	Belgium	8239634
Lm. fermentum	LDTM 7301	Makgeolli	South Korea	5712721

(Table 2) cont.....

LAB Species	Strain	Source Foods	Geographic Location	NCBI BioSample ID
Lm. fermentum	L1	Fermented dough	China	19589998
Lm. fermentum	YL-11	fermented milk	China	10484493
Lm. fermentum	MTCC 25067	Fermented milk	India	6284056
Lb. acidophilus	FSI4	Yogurt	USA	3274004
Weizmannia coagulans	BC01	Thick broad-bean sauce	China	16631050
Lb. johnsonii	BS15	Yogurt	China	4631277
La. casei	MGB0470	Kimchi	South Korea	16657552
Pediococcus acidilactici	PMC65	Kimchi	South Korea	14916503
P. acidilactici	HN9	Traditional Thai fermented pork sausage	Thailand	16191522
P. acidilactici	SRCM102732	Gochujang	South Korea	8707607
Lt. sakei	CBA3614	Kimchi	South Korea	13258981
Lt. sakei	LZ217	Fermented vegetables	China	7305388
Lv. brevis	TMW 1.2108	wheat beer	Germany	4517635
Lv. brevis	NPS-QW-145	Kimchi	Korea	4871683
Lv. brevis	UCCLB95	Beer	Netherlands	8354483
Lv. brevis	NSMJ23	Makgeolli	South Korea	14433864
Lv. brevis	SPC-SNU 70-2	Nuruk	South Korea	19315055
Loigolactobacillus coryniformis	CBA3616	Kimchi	South Korea	11843667
Pediococcus pentosaceus	SRCM102740	Chonggugjang	South Korea	8707615
P. pentosaceus	KCCM 40703	Sake mash	Japan	6447729
P. pentosaceus	wikim20	Kimchi	South korea	4017317
P. pentosaceus	SRCM102734	Doenjang	South Korea	8707609
Leuconostoc citreum	CBA3621	Kimchi	South Korea	11843662
Leu. citreum	37	Yogurts	China	16588752
Lentilactobacillus buchneri	MGB0786	Kimchi	South Korea	12687753
Tetragenococcus halophilus	YJ1	Salty fermented fish sauce	South Korea	13340697
Leuconostoc pseudomesenteroides	CBA3630	Kimchi	South Korea	11843675
Lactiplantibacillus pentosus	BGM48	Olive fermentation	USA	5412454

(Table 2) cont.....

LAB Species	Strain	Source Foods	Geographic Location	NCBI BioSample ID
Lp. pentosus	ZFM222	Fermented vegetables	China	7303442
Lp. pentosus	SLC13	Mustard pickles	Taiwan	7267280
Leu. citreum	CBA3622	Kimchi	South Korea	11843670
Fructilactobacillus sanfranciscensis	LS451	San Francisco sourdough	USA	13111568
Leuconostoc gelidum subsp. gasicomitatum	TMW 2.1619	Meat	Germany	5756392
Lactobacillus amylolyticus	L5	Naturally fermented tofu whey	China	9867391
Lentilactobacillus hilgardii	LMG 07934	Fermented beverages, wine	Germany	14262734
Le. hilgardii	FLUB	Mead	Poland	13567894
Leu.kimchii	NKJ218	Kimchi	South Korea	10887832

Kimchi, a traditional Korean fermented vegetable is prepared with Chinese cabbage as the main ingredient along with ginger, garlic, red pepper powder, fish sauce and radish. LAB and yeast are naturally found in a variety of raw materials, as a result, they play an important role in the fermentation process. *Kimchi* fermentation is quite comparable to other lactic acid fermentation of vegetables, which is due to the wide variety of kimchi products. Initially *Leu. mesenteroides* are the predominant LAB during the *kimchi* fermentation, but with the decrease in pH, other LAB such as *Lp. plantarum, Lv. brevis, Enterococcus faecalis* and *Pediococcus cerevisiae* become dominant [20]. Sauerkraut is another lactic acid fermented vegetable that is frequently consumed in Western countries as it is considered a good source of minerals, polyphenols, vitamins, and glucosinolates (GLS) derived biological active compounds. Sauerkraut is produced under anaerobic conditions by spontaneous fermentation of cabbage following the addition of salt, which is carried out by microbial succession consisting of two phases, a heterofermentative phase dominated by *Leu. mesenteroides* and homofermentative phase dominated by *Lp. plantarum,* respectively. While these bacteria are the most prevalent in sauerkraut fermentation, there are additional microorganisms present in lower quantities which might be essential, namely *Leuconostoc, Lactobacillus, Pediococcus* and *Weissella* species. Similarly, sourdough, a traditional cereal based fermented food product having LAB as the predominant microbe, contains high nutraceutical values because of its positive interaction with yeasts during the fermentation process [22, 23]. According to studies, sourdough has a LAB community with more than 50 species of *Lactobacillus,* being the dominant microbe and more than 20 species of yeast.

Other LAB species, such as *Enterococcus, Lactococcus, Leuconostoc, Pediococcus* and *Weissella* have also been reported in sourdough [21].

Fermented Animal Based Food Product

Several past civilizations as well as in present day, many communities around the world use the process of fermentation to increase the shelf life of meat and dairy products. Numerous fermented food products from animal sources have been obtained by the process of fermentation using safe and harmless microbes, especially LAB. Fermentation with specific LAB strains may result in the toxic removal or anti-nutritional factors, and also degradation of proteins into amino acids and peptides, which results in improving the digestibility of the product [8, 24]. Fermented milk, cheese, fermented sausage, and fermented fish are examples of well-known animal-based fermented foods using LAB. Among LAB, *Lactobacillus, Lactococcus, Leuconostoc, Pediococcus,* and *Streptococcus* are most frequently found in fermented food, either as naturally or as a starter culture.

Yogurt is a fermented milk product which is prepared using *Lb. delbrueckii* subsp. *bulgaricus* and *S. thermophilus,* which have health benefits upon consumption. These bacteria produce lactic acid during yogurt production, which lowers pH and causes milk protein to coagulate. The metabolites produced during the fermentation process, such as non-volatile or volatile acids, carbonyl compounds, and exopolysaccharides, have a significant impact on yogurt quality [25]. Kefir, a traditional fermented milk product, is produced by inoculating kefir grains, which contains numerous microbial species in milk, resulting in sour and yogurt like taste product. Kefir grains have a complex bacterial flora, with *Lactobacillus* species predominating, however *Leuconostoc, Lactococcus, Streptococcus, Acetobacter, Pseudomonas, Acinetobacter,* and other species have been found in various grains [26]. Cheese is a nutrient rich milk traditional product that is consumed worldwide. Cheeses are characterized by firmness, ranging from "soft" to "extra hard," as well as by primary ripening, which ranges from "ripened" to "in brine" [27]. Traditional cheese products contain bioactive compounds such as peptides, fatty acids, and polysaccharides, which have functional properties such as antioxidant, immunomodulatory, antihypertensive, anti-diabetic, and anticancer activity. *Lb. delbrueckii, La. paracasei, Lv. brevis, Lc. lactis, Latilactobacillus curvatus, Lb. helveticus, Lentilactobacillus parabuchneri, La. rhamnosus, Lp. plantarum, Fructilactobacillus fructivorans, and S. thermophilus* are among the LAB species involved in cheese production [8, 28].

Depending on the raw material, processing conditions and starter culture, a wide variety of dry fermented sausages can be found. Appropriate selection of starter cultures with functional properties, especially protease activity, is very much significant in producing fermented sausages, with high antioxidant activity and merry flavour [29]. The most commonly used microorganisms as starter cultures in fermented sausages are *Lactobacillus, Micrococcus, Pediococcus* and *Staphylococcus* species. LAB in traditional fermented sausages includes *Lt. curvatus, Lp. plantarum* and *Lt. sakei* [30]. Fermented fish products are widely consumed in East and South-east Asia, having LAB as the dominant microorganism. Hydrolysis of carbohydrates and lowering the pH are the primary role of LAB in the fermentation process of fish. Production of lactic acid, which lowers pH, helps in the preservation of the fermented fish product. Several beneficial properties of fermented fish products, such as antioxidant, antihypertensive, anticancer, anticoagulant and fibrinolytic activities, have also been reported [31]. *Amylolactobacillus amylophilus, Lo. coryniformis, Lc. lactis, Lp. plantarum,* and *Fructobacillus fructosus* have been reported to be present in fermented fish products [32].

LAB as Starter Cultures and their Selection

Starter culture is defined as preparation or inoculums which contain numerous numbers of variable microbes, which might help in the acceleration, improved controlled process of fermentation and predictability of its end product [8]. In medieval times, fermentation was carried out using inoculums from the previous production and has thus been used as a starter, whereas in modern times, starter cultures are either single or multiple strains which have specific adaptations to a raw material or substrate. Growth of the bacteria naturally present in the raw material resulted in spontaneous fermentation, which was used to make the first fermented foods. Thus, the microbial load and spectrum of the raw material had an impact on the end product's quality. The use of starter cultures in dairy, meat, bread, beverage and other industries has become routine since the nineteenth century and has been a promising method for maintaining the safety and quality of fermented products [8]. The International Dairy Federation (IDF), the European Food and Feed Cultures Association (EFFCA), the European Food Safety Authority (EFSA) and the Food and Drug Administration (FDA) have listed several valuable, technical and safety values of the use of starter cultures [33].

Development of starter culture for lactic acid fermentation depends on the selection of lactic acid producing and stress resistant LAB strains. Lactic acid fermentation results in stressful conditions, such as osmotic stress, temperature change and lactic acid accumulation, leading to quick adaption for microbial proliferation, development of physical properties and taste of end product produced by the microbes [8]. In 1890, the isolation and production of starter cultures were first initiated from sour milk and cheese in Germany and Denmark [33]. The use of starter cultures having certain specific characteristics, such as rapid growth, product yield, higher biomass and also unique organoleptic properties, are employed in food fermentation [34]. Initially, starter cultures having unique properties were isolated, and later random mutagenesis and selection methods were carried out to expand the usage of starter culture in food fermentation. In this respect, conventional methods for microbial food fermentation are still in practice, as genetically modified bacteria have lower acceptance due to technical and safety values. Selection of starter cultures, which are to be employed for the food industries, requires mandatory study of their function and also develops methods which aim to improve and/or increase the functional properties of the culture [33, 35]. Selection of starter cultures depends on the isolation and screening of microbes in small scale food fermentation, and is finalized based on analysis of the functional properties of the food product produced during the fermentation. In recent times, the use of modern tools helps to accurately target diverse genes and metabolic pathways, which are reliable for important functioning of starter cultures. As a result, it helps in the selection of mutant and genetic modified microbes as starter cultures which have better organoleptic properties than that of wild types.

LAB have been used as a starter cultures for food fermentation both traditionally and industrially [36]. LAB generally breaks down the carbon source available in the raw foods to lactic acid along with the decrease in pH during the fermentation process, as a result, it helps in the removal of undesired microbes. It further helps in the increase of texture, flavour, and nutritive and sensory properties which have beneficial health effects. It has also been reported that LAB produces vitamins, sweeteners, sugar polymers, enzymes, aromatic and antimicrobial compounds with probiotic properties [10]. Several LAB present in the gastrointestinal tract can produce B-group vitamins such as folate (vitamin B9), riboflavin (vitamin B2), cobalamin (vitamin B12), and menaquinone (vitamin K2). Furthermore, these vitamins can be found in LAB fermented foods including as cheese, yogurt, and cultured buttermilk. During fermentation, *Bifidobacterium* strains combined with starters such as *Lb. helveticus* MTCC 5463 and *La. rhamnosus* MTCC 5462 in milk, *S. thermophilus* and *Lb. delbrueckii* subsp. *bulgaricus* in yogurt and *S. thermophilus* and/or *Lb. delbrueckii* subsp. *bulgaricus* in skim milk increases the folate content [37].

LAB with different unique characteristics have been isolated from various traditional fermented foods as well as from various raw materials ingested by starter cultures to develop fermented foods [36, 38]. Depending on the nature of the raw material, different LAB are used to produce the desired fermented food product. Several LAB are used as starter cultures such as *Lc. lactis* subsp. *lactis, Lc. lactis* subsp. *lactis var. diacetylactis, Lc. lactis* subsp. *cremoris, Lb. delbrueckii* subsp. *lactis, La. casei, Lb. helveticus* and *Leu. mesenteroides* subsp.

cremoris to produce different varieties of cheese [8, 39]. Fermented milk products are produced using starter cultures, such as *S. thermophilus* and *Lb. delbrueckii* subsp. *bulgaricus* in yogurt, *Lactobacillus kefiranofaciens*, *Lentilactobacillus kefiri* and *Lv. brevis* along with yeasts in kefir, *La. rhamnosus*, *La. casei*, *Lb. acidophilus* and *Lb. johnsonii* in other fermented milk products [38]. Fermented fish products include *Carnobacterium piscicola*, *Companilactobacillus alimentarius* and *Lp. plantarum*, whereas *P. pentosaceus*, *P. acidilactici*, *Lt. sakei* and *Lt. curvatus* are involved in fermented meat products such as sausages [40]. In fermented vegetables, such as olives, sauerkraut, and pickles, species of *Lm. fermentum*, *Lp. plantarum Lp. pentosus*, *Lv. brevis*, *Le. buchneri*, *Leu. mesenteroides*, *P. pentosaceus*, *P. cerevisiae*, *P. acidilactici* and *T. halophilus* are used [41]. Similarly, *C. alimentarius*, *Lactobacillus amylovorus*, *Lv. brevis*, *Companilactobacillus farciminis*, *Lp. plantarum*, *Limosilactobacillus panis*, *Fr. sanfranciscensis*, *Lm. fermentum*, *Limosilactobacillus pontis*, *Lm. reuteri* and *Weissella cibaria* are employed in the fermentation of cereals [42].

Certain gastrointestinal LAB species are able to synthesize B-group vitamins (*i.e.* folate or vitamin B9, riboflavin or vitamin B2, cobalamin or vitamin B12, menaquinone or vitamin K2). Besides, these vitamins exist in fermented foods by LAB such as cheese, yogurt and cultured buttermilk. During fermentation, folate content is increased using strains of *Bifidobacterium* along with starters such as *S. thermophilus* and/or *Lb. delbrueckii* subsp. *bulgaricus* in skim milk, *Lb. helveticus* MTCC 5463 and *La. rhamnosus* MTCC 5462 in milk and *S. thermophilus* and *Lb. delbrueckii* subsp. *bulgaricus* in yogurt [37].

Characterization of LAB Genomes

LAB are widely used in the production of various fermented foods, as a result of which the genome sequencing of LAB is blooming due to its numerous organoleptic properties and beneficial health effects. Genomic information has become vital for the in depth studying of microbes and these data can give an immutable link to the organism that is mostly consistent [43]. The molecular mechanisms behind these characteristics have also been uncovered, and these discoveries have substantially expanded our understanding of LAB. The first reported LAB genome sequence was released in 2001 of *Lc. lactis* spp. *lactis* IL1403. Until now, more than 100 complete genomes of LAB have been sequenced and are available in the public domain, and numerous beneficial traits having industrial importance have been discovered based on the genome analysis of the microorganisms [13]. The LAB genome consists of single circular chromosome having low GC contents and varies in size between 1.3 to 3.3 Mb. The present shape and structure of the LAB genome have been mainly due to some genetic events such as gene duplication, genome reduction and horizontal gene transfer (HGT) [44]. In the last decade, the advent of genomics, functional genomics, and high-throughput technologies has allowed for a comprehensive understanding of the industrial use and metabolic characteristics of LAB, which can aid in the discovery of novel application potentials of LAB worthy of future investigation.

Different Sequencing Approaches

LAB has a number of significant phenotypic traits, which include acidification rate, environmental stress and phage resistance, contribution to texture and flavour development, and probiotic function. These properties are essential for its application and are being constantly investigated. The development of the high through put genome sequencing technologies resulted in a better understanding of LAB physiology, which has evolved significantly. The sequencing approaches used for LAB have been single cell genomics and metagenomics.

Single Cell Genomics

Over the last decade, single cell sequencing methods have become more common. This technique has a lot of promise for a variety of reasons, the most important of which is its ability to discern cellular distinctions within heterogeneous cell populations in any tissue or cell culture. Understanding the development, control, and response to external influences in a population of cells requires determining cell heterogeneity. In bulk sequencing techniques, this natural heterogeneity is pooled and averaged [43]. Traditional sequencing eliminates a lot of information that could reveal more subtle explanations for phenotypes of interest. Many strategies for isolating and sequencing single cells in a cost-effective and high-throughput way have been developed [45]. Microfluidics and Fluorescent Activated Cell Sorting (FACS) are the most widely used technologies to date. FACS relies on fluorescent cell tagging and isolation by taking advantage of the charged nature of a fluorescently marked cell, whereas microfluidics is concerned with the precise mixture of oil, surfactants, and cells in order to produce a droplet containing a single cell [46]. These approaches are employed in a wide range of fields and are ideal for single cell sequencing. Furthermore, these approaches have been modified to include cell lysis and the incorporation of sequencing materials into the droplets encapsulating the cell components.

Single cell sequencing approach has been currently limited to human cells, and due to not having suitable lysing protocols for bacterial cells, there are no LAB studies which use high throughput analysis. Despite this drawback, studies have demonstrated potential for LAB research in this field based on the proof of principle and also other approaches have been reported which are directly relevant to LAB. Moreover, the rapid advancement of isolation methods, lysing techniques, and sequencing depth will result in a more reliable platform for LAB targeted single cell investigation in the future.

Metagenomics

Metagenomics provides community based genome sequences of diverse species simultaneously and also provides an overview of species abundance in the microbiome, characterizing the common metabolic pathways which are available in the ecosystem. The Metagenomics approach allows to identify microorganisms which are able to enhance sensory quality, increase bioavailability of nutrients, improve safety, biopreservative effects of fermented foods and provide beneficial effects to human. Using the metagenomic data, contribution of LAB to functional processes and gene pool can be identified, allowing us to understand the potential role of LAB in the community. The first metagenomic study based on sequencing was performed for kimchi, a traditional fermented food which revealed the presence of prevalent microbial populations such as *Lt. sakei* and *Leu. mesenteroides* [47].

Before evaluating single-strain physiology and genetic background, it is required to first characterize the starter composition that is to specify the strains it contains and the associated numbers. Metagenomics study gives culture-independent insights into species composition and dynamics. The systematic decoding of genetic information in an undefined starter community through metagenomics analysis improves the studies on starter LAB, like high-resolution surveillance of microbial community dynamics or the development of community-based genome-scale metabolic models. Thus, it will help with the understanding of growth, metabolism, and physiology during mixed culture preparation, fermented food making, and ripening. Several novel data analysis tools based on reference-mapping approaches have recently been developed and evaluated for strain-level metagenomics data analysis. The development of these new tools has caught the interest of the research community, especially in the field of LAB, as it is now able to provide a more in-depth knowledge of the starter culture and its significance to the quality of fermented products [48].

PRACTICAL INSIGHTS INTO LAB GENOMES

For centuries, LAB have been widely employed in the various food industries, particularly in the production of fermented foods. Due to its GRAS status and broad potential application in the fermentation industry, several beneficial properties of LAB, such as β-galactosidase, antimicrobial agents, bile salt hydrolase, exopolysaccharide and GABA-producing LAB have attracted a lot of interest in recent years (Table **3**). In this section, we briefly discuss the genes associated with the above mentioned properties and its expression to enhance the beneficial properties of LAB.

Table 3. Details of genes and their encoded products in selected LAB species.

LAB Species	Gene	Encoded Products	References
Lb. acidophilus	*lacA*	β-galactosidase	[101]
Leu. mesenteroides subsp. mesenteroides	*lacZ*	β-galactosidase	[87]
Lentilactobacillus parakefiri	B8W98_01810, C5L28_000990	β-galactosidase	[13]
La. paracasei	LSEI_0384	β-galactosidase	[87]
Lm. *reuteri*	*lacM*	β-galactosidase	[102]
La. rhamnosus	*uidA*	β-galactosidase	[103]
O. oeni	*lacZ*	β-galactosidase	[87]
Lc. lactis subsp. lactis	*lctA*	Bacteriocins	[104]
Leu. gelidum	*lcnA*	Bacteriocins	[105]
Lc. lactis subsp. lactis	*nisZ*	Bacteriocins	[106]
Lt. sakei	*sppA*	Bacteriocins	[107]
Lactiplantibacillus paraplantarum	*plnA*	Bacteriocins	[108]
Lb. acidophilus	*bshA, bshB*	Bile salt hydrolase	[109]
Lg. salivarius	*bsh1*	Bile salt hydrolase	[110]
Lc. lactis	*epsB, espC*	Exopolysaccharide	[111]
S. thermophilus	*espA, epsB, espC, epsD*	Exopolysaccharide	[112]
Leu. pseudomesenteroides	*epsB*	Exopolysaccharide	[111]
Lc. lactis subsp. cremoris	*repC*	Exopolysaccharide	[113]
Lc. lactis subsp. cremoris	*gadC*	GABA	[114]
Lc. lactis subsp. lactis	*gadC*	GABA	[115]
Lc. lactis subsp. lactis KF147	*gadC*	GABA	[116]
Lactobacillus antri DSM 16041	*gadC*	GABA	[117]
Lc. lactis subsp. cremoris KW2	*gadC*	GABA	[118]
Lc. lactis subsp. lactis IO-1	*gadC*	GABA	[119]
Lm. *reuteri*	*gadC*	GABA	[102]
Lc. lactis subsp. cremoris MG1363	*gadB*	GABA	[74]
Lc. lactis subsp. lactis IL1403	*gadB*	GABA	[12]

β-galactosidase

β-galactosidase, which is a vital enzyme for both the human beings and the food industry, catalyses the hydrolysis and transgalactosylation of β-galactosides like lactose. In the dairy industry, β-galactosidase is commonly utilized to improve lactose digestion, solubility, and sweetness [14]. Transgalactosylation is an effective approach for converting lactose into galactooligosaccharides (GOS), which are considered probiotics because they stimulate the growth of beneficial microbes in the intestine. Also, the formation of colourful compounds by this enzyme during a chemical reaction has attracted a quite good attention of molecular biologists. Lactose intolerance has become a common problem in humans, characterized by a reduced ability to digest lactose due to a deficiency of β-galactosidase on the brush edge of the duodenum [14]. Individuals with lactose intolerance must limit their intake of lactose and avoid drinking milk, as a result, it may also limit their intake of key nutrients such as proteins and calcium.

The use of lactose-reduced milk products or the supplementation of the diet with exogenous β-galactosidase is a common technique to control lactose intolerance [49]. Unfortunately, eradicating lactose alters the quality and flavour of milk products and does not completely remove lactose. β-galactosidase is also synthesized by some intestinal microflora and LAB in products like yogurt, which can help to reduce lactose intolerance. As a result, for enhancing β-galactosidase expression in food-grade, LAB is a potential alternative lactose intolerance management technique as several LAB has been reported to synthesize β-galactosidase.

Until now, β-galactosidases have been classified into four glycoside hydrolase (GH) groups based on amino acid composition, which include GH1, GH2, GH35, and GH42 [50]. Although most β-galactosidases are expressed by a single gene, current research has revealed that some GH2 β-galactosidases from LAB are composed of two subunits (heterodimer), first discovered in *Leuconostoc lactis* [51]. Some LAB, such as *Lb. acidophilus*, have both GH2 and GH4 lactase, which are denoted as LacLM and LacZ, respectively, unlike other microbes that contain only one group of β-galactosidase [52]. The development of the two subunits was attributed to an internal loss in an ancestral β-galactosidase gene, according to DNA and protein sequence alignments [51]. A similar phenomenon has also been observed in *Lb. kefiranofaciens*, in which the *lacLM*, belonging to the GH2 family, encodes heterodimeric β-galactosidase, consisting of two overlapping genes *lacL* and *lacM,* encoding small and large subunits with molecular mass of 35 and 75 kDa, respectively. Whereas the *LacZ* is present in *Lb. kefiranofaciens,* belonging to the GH42 family, encodes a single polypeptide chain with an estimated molecular mass of 76 kDa. It has also been observed that when the *lacLM* and *lacZ* genes of *Lb. kefiranofaciens* ZW3 strain isolated from kefir grains expressed in *E. coli*, the small subunit may have played a significant role in the structural stability of the enzyme [53]. In a study, it has been reported that one strain of *Lm. reuteri* had high β-galactosidase activity along with substantial transferase activity, and the heterodimeric β-galactosidase is encoded by *lacL* and *lacM* overlapping genes [52].

Antimicrobial Agents

LAB produce small heat stable peptides, which are known as bacteriocins that have antimicrobial properties. Bacteriocins have diverse applications, and significant research is done to discover new bacteriocins, having novel roles for these compounds. Bacteriocins have been believed. With the rise in demand for shelf stable foods which are minimally processed, novel bacteriocins can be used as adjuvants in food products for further protection against bacterial growth [54]. Bacteriocins production depends on the microbial strain and conditions for its growth. Newly synthesized bacteriocins have an N-terminal leader, which is modified by amino acids or proteins encoded by bacteriocin gene cluster before being exported out of the cell [55]. Genes that encode the production of bacteriocins and immunity are organized in operon clusters, and can be located on chromosomes that are associated with plasmids or transposons [16]. In seven LAB genomes, bacteriocin gene clusters have been identified, including the discovery of incomplete bacteriocins production pathways from non-bacteriocin producers [54].

The primary pathways for Class I bacteriocins, which are lantibiotic, production, can be described using the well-known nisin pathway, with minor changes for non-lantibiotic bacteriocins because they do not require the inclusion of unique amino acids [16]. *Lactococcus* spp. produce lantibiotics, that binds to lipid-II, a crucial precursor of peptidoglycan synthesis, forming a complex enabling to suppress cell wall synthesis. The N-terminus is the important domain as it is linked to the peptide's C-terminal end by the hinge region, containing amino acids 20, 21, and 22, thus allowing pore formation in the target membrane [56]. Plantaricin, a class II bacteriocin produced by *Lp. plantarum*, can be chromosomally or plasmid-encoded. Plantaricin 423 is encoded by a plasmid, whereas plantaricin ST31 is encoded by a chromosome. Class II bacteriocin production genes are grouped into operon clusters and typically include a specialised immunity gene, a structural gene encoding the prepeptide, an ABC-transporter gene for membrane translocation, and an accessory protein gene for bacteriocin export, and sometimes regulatory genes are also found [57, 58]. Class II bacteriocins are biosynthesized as an inactive prepeptide with an N-terminal leader peptide and an unique double-glycine proteolytic processing site, whereas

Class IIc bacteriocins contain a sec-type N-terminal signal sequence and are processed and released *via* the general secretary pathway. The structural genes, immunity gene, accessory protein, and ABC transporter are all encoded by four ORFs (plaABCD) in the plantaricin 423 encoding area, which is identical to the pediocinPA-1 encoding region. Plantaricin 423'splaC and plaD genes are 99 percent identical to pedC and pedD from pediocin PA-1 [16]. *Lactobacillus* spp. and *Enterococcus* spp. strains are the most common producers of Class III bacteriocins, which are large peptides and heat-unstable proteins. Class III bacteriocins aren't particularly interesting because they can't withstand thermal processing and haven't been thoroughly studied [56].

Bile Salt Hydrolase

Bile salt hydrolase, also known as choloylglycine hydrolase, is a member of the N-terminal nucleophilic hydrolase superfamily, which catalyses the hydrolysis of conjugated bile salts, resulting in free bile salts and amino acid residues [59]. Through dehydroxylation, dehydrogenation, and sulfation, this process opens a larger pathway of bile acid alteration encoded by the gut bacteria, which yields secondary and tertiary forms. Bile acids act as signaling molecules in the intestine, controlling its biosynthesis, lipid metabolism, and local mucosal defenses, therefore having potential host physiological influence [18]. When bile salt hydrolase is expressed in the intestinal lumen, it leads to the reduction of conjugated bile salts while increasing the levels of free bile salts, which are less soluble and absorbed less effectively unlike to their conjugated counterparts. As a result, bile salt reabsorption and recycling are reduced, causing an increase in the liver's *de novo* bile salt synthesis from cholesterol [60]. Bile salt hydrolase enzymes may also influence membrane potential, fluidity, and tensile strength by promoting cholesterol integration into bacterial membranes, and are thought to promote gut bacteria viability by increasing bile salt tolerance and sensitivity to host defensins. Bile salt hydrolase enzymes are considered to be buffer shock proteins that help *Lactobacilli* survive in the GI tract and a significant association between microbial habitat and bile salt hydrolase activity has been shown to support the findings [61].

Bile salt hydrolase activity has been observed in a wide range of gut-associated bacterial phyla, including Actinobacteria, Bacteroidetes, and Firmicutes, with Firmicutes bacteria having the highest bile salt hydrolase gene abundance in the human gut [62]. LAB constitutes a large fraction of the Firmicutes bacteria found in the human gut, and bile salt hydrolase activity has been observed in various lactic acid bacterial genera, which include *Bifidobacterium*, *Enterococcus*, *Lactobacillus* and *Pediococcus*. *Lb. johnsonii* PF01 and *Lg. salivarius* LMG14476 was observed to have three and two bile salt hydrolase enzymes respectively which had substantially varied catalytic efficiency and substrate preference [63].

Exoploysaccharides

Exopolysaccharides (EPS) contribute to distinct physical properties such enhanced texture and rheology, which have been studied for numerous application in the dairy industry. EPS has been widely produced by LAB in fermentation food to extend shelf-life and preserve flavour by influencing syneresis, viscosity, and sensory qualities [64, 65]. Extracellular polysaccharides produced by microbes are frequently employed in the cosmetics, food, and pharmaceutical industries. EPS produced by LAB have gained substantial interest due to their potential use in the food sector. Several LAB species, such as *Fructilactobacillus*, *Lacticaseibacillus*, *Lactiplantibacillus*, *Lactobacillus*, *Lactococcus*, *Latilactobacillus*, *Lentilactobacillus*, *Leuconostoc*, *Limosilactobacillus*, *Pediococcus*, *Streptococcus*, and *Weissella* are known to synthesize numerous EPS [9].

Two types of EPS are synthesized by LAB, which are homopolysaccharides (HoPS) and heteropolysaccharides (HePS). The majority of EPS produced by LAB are HePS. There are two major types of enzymes involved in the synthesis of EPS by LAB. Proteins in the first group are required for the synthesis of basic sugar nucleotides, which are utilized by other cell pathways. The other group contains EPS-specific enzymes that regulate the entire process, such as glycosyl- and acetyl-transferases, which use monomeric molecules as the glycosyl donor during synthesis, or enzymes that are responsible for polymerization and export [66]. Specific genes control the activity of EPS-specific enzymes, such as in thermophilic LAB, the genes encoding EPS biosynthesis are usually grouped in a cluster and are chromosomal, whereas they can also be found on plasmids in mesophilic LAB [67]. The *eps* genes are arranged in a cluster with an operon structure having all of the genes orienting in the same direction and are transcribed as a single mRNA, with promoter sequence in front. The *eps* genes are responsible for four distinctive functions, which include regulating EPS production, determining chain length, biosynthesis of repeating units, and polymerization and export. A typical *eps* gene cluster in LAB includes five highly conserved genes, *eps*A, *eps*B, *eps*C, *eps*D, and *eps*E, as well as a more variable region that includes the polymerase *wzy*, the flippase *wzx*, and glucosyltransferases. Multiple EPS clusters have been identified in *Lp. plantarum*, whereas the *eps* gene cluster in *Lactobacilli* is more sophisticated than in other species such as *Lc. lactis* and *S. thermophilus* [68].

Gamma-aminobutyric Acid

Gamma-aminobutyric acid (GABA) is a non-protein amino acid that is formed by the -decarboxylation of L-glutamic acid, which is catalysed by glutamate decarboxylase. GABA has antianxiety, antidepressant, and antihypertensive effects in addition to acting as an inhibitory neurotransmitter [8]. Furthermore, GABA consumption may suppress cancer cell growth, can be administered in the treatment of Parkinson's and Alzheimer's disease, also can regulate thyroid hormone levels, and strengthen the immune system [69]. GABA is produced by LAB during fermentation as a defensive response to maintain its viability in the acidic environment [70]. The ability of different LAB strains to produce GABA varies and is affected by different fermentation parameters or additives such as cultivation time, temperature, pH, PLP addition, and medium composition. *Enterococcus avium*, *Levilactobacillus namurensis*, *P. pentosaceus*, *Lb. bulgaricus*, *Le. buchneri*, *Lb. helveticus*, *La. paracasei*, *Lc. lactis*, *Lv. brevis*, *Levilactobacillus zymae* and *Companilactobacillus futsaii* have been reported to be capable of producing GABA [17].

The conversion of glutamate to γ-aminobutyric acid is mediated by glutamate decarboxylase, which catalyses the irreversible α-decarboxylation of glutamate. GAD catalyses the decarboxylation of L-glutamate with the cofactor pyridoxal-5--phosphate (PLP), resulting in the synthesis of GABA and the release of CO_2 as a byproduct [71]. The glutamate-dependent acid-resistance system (GDAR) protects against an acidic environment allowing LAB to recognize and tolerate acid stress which is critical for successful colonization of the gastrointestinal tract (GIT) and survival in acidic conditions during the fermentation. The GDAR system is made up of two homologous inducible glutamate decarboxylases, GadA and GadB, as well as the glutamate/-aminobutyrate antiporter GadC [72]. GABA-degradative enzyme, GABA aminotransferase (GABA-AT), degrades GABA to succinic semialdehyde, which is finally converted to succinic acid by succinate semialdehyde dehydrogenase and enters the TCA cycle. The GABA-AT encoding gene *gad*T has been found in *Limosilactobacillus gastricus*, *Lm. fermentum*, *Lm. pontis*, *Limosilactobacillus mucosae*, *Limosilactobacillus oris*, *Lp. plantarum*, *Lm. reuteri*, *Limosilactobacillus frumenti*, *Lactobacillus gorilla*, *Leu. mesenteroides subsp. mesenteroid*, *Secundilactobacillus similis*, *Limosilactobacillus vaginalis*, *Leu. citreum*, *Leuconostoc kimchi* and *O. oeni* species [73].

GAD is a crucial enzyme in the bioconversion of GABA that is found in the cytoplasm and can be synthesised by strains carrying the GAD gene. The majority of GABA-producing LAB species have a GAD-encoding gene. The GAD systems of most LAB species are present on the chromosome, whereas *Lt. sakei* WiKim0074's is present on its plasmid. Some species have the glutamate-tRNA ligase gene gltX upstream of the gadB-gadC operon. In *Lc. lactis*, glutamate synthases producing genes gltB and gltC are found near the *gadB-gadC* gene [74]. Interestingly, *E. avium* 352 has three GADs, whereas the majority of *Lv. brevis* strains have two unique GAD producing genes [75]. A high GABA-producing *Lv. brevis* CD0817 was found to have just a gadB-gadC operon and no gadA gene [76]. The gadC gene encodes a GABA antiporter, which is involved in both GABA export and glutamate import. Most of LAB species have GABA antiporter, except *Lm. fermentum,* which possesses a GAD not accompanied by a GABA antiporter, whereas in the case of *Lm. reuteri*, it has two GABA antiporters [77].

APPLICATIONS OF GENOME EDITING IN LAB

Genome editing of lactic acid bacterial strains has a lot of potential for developing functional foods with a favourable influence on the food industry (Fig. **2**). *Lc. lactis*, for example, is widely utilized in food fermentations and is a major industrial bacterial culture. Its economic significance, as well as a long history of safe use, has driven the development of genetic and metabolic engineering technology. Another important LAB species is *S. thermophilus*, which has probiotic potential and is widely employed in the production of fermented dairy foods such as yogurt and several cheese variants.

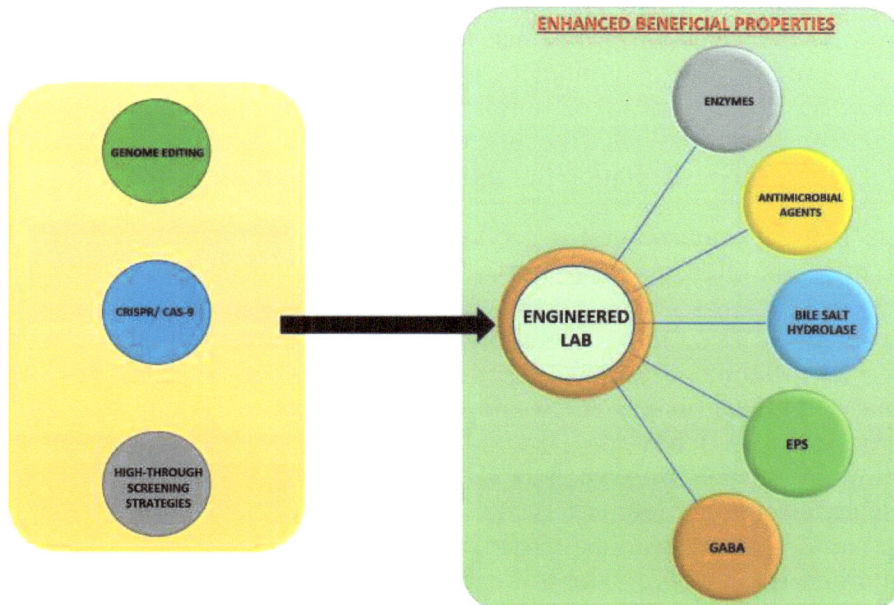

Fig. (2). Genetic engineering and screening of LAB to produce enhanced beneficial properties. EPS: Exopolysaccharide; GABA: β-galactosidase.

Since health-promoting benefits are strain dependent, proper identification of probiotic microorganisms is required and has become essential to use existing tools to identify the microorganisms [78]. As a result, genomic techniques are the most appropriate methodology for illustrating and typing genetic relationships between microbial strains designed for probiotic purposes. *Lactobacilli* are suitable candidates which can be engineered for delivery of therapeutic proteins and peptides such as cytokines, antigens, antibodies, and nanobodies due to their safety and intrinsic health benefits [79]. Genome editing can be used to fix naturally inactivated genes or to activate silent gene clusters, genome editing can accomplish gene replacement, point mutation, or deletion. For example, activation of a silent promoter by point mutation and single-base-pair insertion in *S. thermophilus* strain CNRZ 302 restored galactose fermentation [80]. Furthermore, by replacing the nonsense mutation in the putative two-component system response regulator, bacteriocin production in *Lb. gasseri* DSM 14869 could be restored [81].

LAB must survive a variety of extreme conditions, such as oxidative and osmotic stress, acid and bile, pathogens, and the host immunological response, in industrial environments or in the gastrointestinal tract. In *Bifidobacterium animalis* subsp. *lactis* and *Lb. johnsonii*, spontaneous mutations of the hypothesised membrane-anchored protein Balat_1410 and the putative tyrosine kinase EpsC altered EPS characteristics, have resulted in cells that were resistant to gastrointestinal stress [82]. Further, a mutant strain with galactose-consuming ability was produced by a spontaneous mutation in the galKTEM promoter of *S. thermophilus*. The insertion of single nucleotide mutations using developing genome editing technologies, like as the CRISPR–Cas systems, would surely be faster than spontaneous mutation through successive cultures.

With the help of genome editing, LAB are also employed as vaccine carriers because they can elicit mucosal and systemic immune responses while avoiding the hazards associated with conventional attenuated live pathogens [83]. The use of engineered *Lc. lactis* that secrete interleukin-10 (IL-10) to treat colitis-induced animals with inflammatory bowel disease (IBD) have been reported [84]. Further, additional cytokines, such as IL-12 and IL-6, have been generated in *Lc. lactis* for IBD treatment. LAB has also been developed as a cell factory for allergen production and distribution, such as *Lp. plantarum* NCL21 was used to produce Japanese cedar pollen allergen Cry j1, *Lc. lactis* CHW9 which have been used to produce Ara 2, a major peanut allergen, and *Lc. lactis* NZ9800 was used to deliver birch allergen Bet-v1 [82]. The utilization of engineered LAB in the functional food industry has a lot of potentials, though further studies are to be done.

FUTURE DIRECTIONS AND CONCLUSION

Ethnic and traditional fermented food products are essential as they play prominent roles in the traditions, histories, and social and economics. LAB fermented food provides health promoting benefits upon consumption. LAB starter culture modification can help to improve the hygiene, safety, and quality of fermented products. As commercial LAB starter cultures can considerably improve fermented food products, but the present day molecular and technical development provide far improved and faster prospects having increased accuracy and sensitivity. As there is a global demand for functional foods and fermentation is a low-cost and environmentally friendly biotechnological method of producing them. For many years, the application of probiotic LAB has garnered a lot of attention, resulting in the production of a variety of food products with various functional properties and health benefits.

Moreover, with the increasing amount of published genomes data, it provides an unparalleled potential to identify significant LAB features. The genetics of LAB gives vital insights into their physiology, metabolism, and future application in the food sector. Furthermore, as the knowledge of the metabolic properties of LAB advances, LAB could be employed as cell factories for the manufacture of a wide range of compounds. More LAB genomes will be explored as next generation sequencing technology progresses, contributing to improvements in physiology, genetics, and application of LAB. Furthermore, researchers will gain a better grasp of the biosynthetic and metabolic capacities, stress response, and probiotic action mechanisms of LAB. Though, the primary challenge for this technique is resolution, and with the increasing precision of sequencing in the near future, it will greatly facilitate the study of microbial consortia at the strain level.

Also, the production of metabolites such as β-galactosidase, antimicrobial agents, bile salt hydrolase, exopolysaccharide and GABA during lactic acid fermentation is quite beneficial in improving the nutritional values of the fermented food product. This has also led to increase shelf life and sensory properties, as of which these metabolites have attracted attention in the recent past to enhance the organoleptic properties of fermented food end product. Further, genome editing may be utilized to develop unique LAB strains with specific traits for particular application, which is currently used for production purposes. Also, the strain improvement can be done using genome editing, and the relation with its host can provide knowledge for further modification, which can be generally accepted. Moreover, other advanced genome editing technologies for a broader range of LAB are needed to achieve such advancements, such as multiplex genome silencing and editing.

ACKNOWLEDGEMENTS

The authors would like to acknowledge the Department of Biotechnology (DBT), Government of India, for financial support. We thank ED, NABI and CEO, CIAB, for kind support and encouragement.

REFERENCES

[1]   Ghosh T, Beniwal A, Semwal A, Navani NK. Mechanistic insights into probiotic properties of lactic acid bacteria associated with ethnic fermented dairy products. Front Microbiol 2019; 10(MAR): 502.
[http://dx.doi.org/10.3389/fmicb.2019.00502] [PMID: 30972037]

[2]   Kumari R, Sanjukta S, Sahoo D, Rai AK. Functional peptides in asian protein rich fermented foods: Production and health benefits. Syst Microbiol Biomanufacturing 2021; 1: 3.

[3]   Sarkar P, Abedin MM, Singh SP, Pandey A, Rai AK. Microbial production and transformation of polyphenols. Curr Dev Biotechnol Bioeng. 2022; pp. 189-208.

[4]   Sanjukta S, Sahoo D, Rai AK. Fermentation of black soybean with *Bacillus* spp. for the production of kinema: Changes in antioxidant potential on fermentation and gastrointestinal digestion. J Food Sci Technol 2021; 2021: 1-9.
[PMID: 35250060]

[5]   Chourasia R, Padhi S, Chiring Phukon L, Abedin MM, Singh SP, Rai AK. A potential peptide from soy cheese produced using *Lactobacillus delbrueckii* WS4 for effective inhibition of SARS-CoV-2 main protease and S1 glycoprotein. Front Mol Biosci 2020; 7: 601753.
[http://dx.doi.org/10.3389/fmolb.2020.601753] [PMID: 33363209]

[6]   Chourasia R, Padhi S, Phukon LC, *et al.* Peptide candidates for the development of therapeutics and vaccines against β-coronavirus infection. Bioengineered 2022; 13(4): 9435-54.
[http://dx.doi.org/10.1080/21655979.2022.2060453]

[7]   Padhi S, Chourasia R, Kumari M, Singh SP, Rai AK. Production and characterization of bioactive peptides from rice beans using *Bacillus subtilis*. Bioresour Technol 2022; 351: 126932.
[http://dx.doi.org/10.1016/j.biortech.2022.126932] [PMID: 35248709]

[8]   Chourasia R, Abedin MM, Chiring Phukon L, Sahoo D, Singh SP, Rai AK. Biotechnological approaches for the production of designer cheese with improved functionality. Compr Rev Food Sci Food Saf 2021; 20(1): 960-79.
[http://dx.doi.org/10.1111/1541-4337.12680] [PMID: 33325160]

[9]   Zheng J, Wittouck S, Salvetti E, *et al.* A taxonomic note on the genus *Lactobacillus*: Description of 23 novel genera, emended description of the genus *Lactobacillus beijerinck* 1901, and union of *Lactobacillaceae* and *Leuconostocaceae*. Int J Syst Evol Microbiol 2020; 70(4): 2782-858.
[http://dx.doi.org/10.1099/ijsem.0.004107] [PMID: 32293557]

[10]   Landete JM. A review of food-grade vectors in lactic acid bacteria: From the laboratory to their application. Crit Rev Biotechnol 2017; 37(3): 296-308.
[http://dx.doi.org/10.3109/07388551.2016.1144044] [PMID: 26918754]

[11]   Rai AK, Pandey A, Sahoo D. Biotechnological potential of yeasts in functional food industry. Trends Food Sci Technol 2018; 2019(83): 129-37.

[12]   Bolotin A, Wincker P, Mauger S, *et al.* The complete genome sequence of the lactic acid bacterium *Lactococcus lactis* ssp. lactis IL1403. Genome Res 2001; 11(5): 731-53.
[http://dx.doi.org/10.1101/gr.169701] [PMID: 11337471]

[13]   Buron-Moles G, Chailyan A, Dolejs I, Forster J, Mikš MH. Uncovering carbohydrate metabolism through a genotype-phenotype association study of 56 lactic acid bacteria genomes. Appl Microbiol Biotechnol 2019; 103(7): 3135-52.

[http://dx.doi.org/10.1007/s00253-019-09701-6] [PMID: 30830251]

[14]   Xavier JR, Ramana KV, Sharma RK. β-galactosidase: Biotechnological applications in food processing. J Food Biochem 2018; 42(5): e12564.
[http://dx.doi.org/10.1111/jfbc.12564]

[15]   Thakur M, Amit R, Sudhir K, Singh P. An acid-tolerant and cold-active β-galactosidase potentially suitable to process milk and whey samples. Appl Microbiol Biotechnol 2022; 2022: 1-12.

[16]   Mokoena MP. Lactic acid bacteria and their bacteriocins: classification, biosynthesis and applications against uropathogens: A Mini-Review. Mol 2017; 22: 1255.

[17]   Yogeswara IBA, Maneerat S, Haltrich D. Glutamate decarboxylase from lactic acid bacteria: A key enzyme in GABA synthesis. Microorg 2020; 8(12): 1923.

[18]   Ru X, Zhang CC, Yuan YH, Yue TL, Guo CF. Bile salt hydrolase activity is present in nonintestinal lactic acid bacteria at an intermediate level. Appl Microbiol Biotechnol 2018; 103(2): 893-902.

[19]   Mao B, Yan S. Lactic acid bacteria and fermented fruits and vegetables. Lact Acid Bact. 2019; pp. 181-209.
[http://dx.doi.org/10.1007/978-981-13-7283-4_7]

[20]   Fan L, Hansen LT. Fermentation and biopreservation of plant-based foods with lactic acid bacteria. 2nd. Handb Plant-Based Fermented Food Beverage Technol 2012; pp. 23-34.
[http://dx.doi.org/10.1201/b12055-5]

[21]   Torres S, Verón H, Contreras L, Isla MI. An overview of plant-autochthonous microorganisms and fermented vegetable foods. Food Sci Hum Wellness 2020; 9(2): 112-23.
[http://dx.doi.org/10.1016/j.fshw.2020.02.006]

[22]   Sakandar HA, Hussain R, Kubow S, Sadiq FA, Huang W, Imran M. Sourdough bread: A contemporary cereal fermented product. J Food Process Preserv 2019; 43(3): e13883.
[http://dx.doi.org/10.1111/jfpp.13883]

[23]   Phukon LC, Singh SP, Pandey A, Rai AK. Microbial bioprocesses for production of nutraceuticals and functional foods. Curr Dev Biotechnol Bioeng 2022; 1-29.
[http://dx.doi.org/10.1016/B978-0-12-823506-5.00001-1]

[24]   Abedin MM, Phukon LC, Chourasia R, Sharma S, Sahoo D, Rai AK. Soybean-derived bioactive peptides and their health benefits. Phytochem Soybeans 2022; 455-74.
[http://dx.doi.org/10.1201/9781003030294-20]

[25]   Nagaoka S. Yogurt production. Methods Mol Biol 2019; 1887: 45-54.
[http://dx.doi.org/10.1007/978-1-4939-8907-2_5] [PMID: 30506248]

[26]   Walsh AM, Crispie F, Kilcawley K, *et al.* Microbial succession and flavor production in the fermented dairy beverage kefir. mSystems 2016; 1(5): e00052-16.
[http://dx.doi.org/10.1128/mSystems.00052-16] [PMID: 27822552]

[27]   Abedin MM, Chourasia R, Chiring Phukon L, Singh SP, Kumar Rai A. Characterization of ACE inhibitory and antioxidant peptides in yak and cow milk hard *chhurpi* cheese of the Sikkim Himalayan region. Food Chem X 2022; 13: 100231.
[http://dx.doi.org/10.1016/j.fochx.2022.100231] [PMID: 35499015]

[28]   Chourasia R, Kumari R, Singh SP, Sahoo D, Rai AK. Characterization of native lactic acid bacteria from traditionally fermented chhurpi of Sikkim Himalayan region for the production of chhurpi cheese with enhanced antioxidant effect. Lebensm Wiss Technol 2022; 154: 112801.
[http://dx.doi.org/10.1016/j.lwt.2021.112801]

[29]   Cao CC, Feng MQ, Sun J, Xu XL, Zhou GH. Screening of lactic acid bacteria with high protease activity from fermented sausages and antioxidant activity assessment of its fermented sausages CyTA - Journal of Food 2019; 17(1): 347-54. Available from: http://mc.manuscriptcentral.com/tcyt
[http://dx.doi.org/10.1080/19476337.2019.1583687]

[30] Najjari A, Boumaiza M, Jaballah S, Boudabous A, Ouzari HI. Application of isolated *Lactobacillus sakei* and *Staphylococcus xylosus* strains as a probiotic starter culture during the industrial manufacture of Tunisian dry-fermented sausages. Food Sci Nutr 2020; 8(8): 4172-84.
[http://dx.doi.org/10.1002/fsn3.1711] [PMID: 32884698]

[31] Zang J, Xu Y, Xia W, Regenstein JM. Quality, functionality, and microbiology of fermented fish: A review. Crit Rev Food Sci Nutr 2019; 60(11): 1-15.

[32] Chowdhury N, Goswami G, Hazarika S, Sharma Pathak S, Barooah M. Microbial dynamics and nutritional status of namsing: A traditional fermented fish product of mishing community of assam. Proceedings of the National Academy of Sciences, India Section B: Biological Sciences 2018; 89(3): 1027-38.

[33] Kandasamy S, Kavitake D, Shetty PH. Lactic acid bacteria and yeasts as starter cultures for fermented foods and their role in commercialization of fermented foods. Innov Technol Fermented Food Beverage Ind 2018; 25-52.
[http://dx.doi.org/10.1007/978-3-319-74820-7_2]

[34] Kavitake D, Kandasamy S, Devi PB, Shetty PH. Recent developments on encapsulation of lactic acid bacteria as potential starter culture in fermented foods : A review. Food Biosci 2017; 2018(21): 34-44.

[35] Chourasia R, Phukon CL, Abedin MM, Sahoo D, Rai AK. Microbial transformation during gut fermentation. Bioact Compd Fermented Foods 2021; 365-402.

[36] Lasrado LD, Rai AK. Recovery of nutraceuticals from agri-food industry waste by lactic acid fermentation. Energy, Environ Sustain 2018; 185-203.
[http://dx.doi.org/10.1007/978-981-10-7434-9_11]

[37] Saubade F, Hemery YM, Guyot JP, Humblot C. Lactic acid fermentation as a tool for increasing the folate content of foods. Critical Reviews in Food Science and Nutrition 2017; 57(18): 3894-910.
[http://dx.doi.org/10.1080/10408398.2016.1192986]

[38] Mohammadi R, Sohrabvandi S, Mohammad Mortazavian A. The starter culture characteristics of probiotic microorganisms in fermented milks. Eng Life Sci 2012; 12(4): 399-409.
[http://dx.doi.org/10.1002/elsc.201100125]

[39] Chourasia R, Phukon LC, Abedin MM, Sahoo D, Rai AK. Production and characterization of bioactive peptides in novel functional soybean chhurpi produced using *Lactobacillus delbrueckii* WS4. Food Chem 2022; 387: 132889.
[http://dx.doi.org/10.1016/j.foodchem.2022.132889]

[40] Kopermsub P, Yunchalard S. Identification of lactic acid bacteria associated with the production of plaa-som, a traditional fermented fish product of Thailand. Int J Food Microbiol 2010; 138(3): 200-4.
[http://dx.doi.org/10.1016/j.ijfoodmicro.2010.01.024] [PMID: 20167386]

[41] Bağder Elmacı S, Tokatlı M, Dursun D, Özçelik F, Şanlıbaba P. Phenotypic and genotypic identification of lactic acid bacteria isolated from traditional pickles of the Çubuk region in Turkey. Folia Microbiol 2015; 60(3): 241-51.
[http://dx.doi.org/10.1007/s12223-014-0363-x] [PMID: 25404550]

[42] Katina K, Poutanen K. Nutritional aspects of cereal fermentation with lactic acid bacteria and yeast. Handb Sourdough Biotechnol 2013; 229-44.
[http://dx.doi.org/10.1007/978-1-4614-5425-0_9]

[43] O'Donnell ST, Ross RP, Stanton C. The progress of multi-omics technologies: Determining function in lactic acid bacteria using a systems level approach. Front Microbiol 2020; 10(January): 3084.
[http://dx.doi.org/10.3389/fmicb.2019.03084] [PMID: 32047482]

[44] Mayo B, Sinderen D, Ventura M. Genome analysis of food grade lactic Acid-producing bacteria: from basics to applications. Curr Genomics 2008; 9(3): 169-83.
[http://dx.doi.org/10.2174/138920208784340731] [PMID: 19440514]

[45]   Hwang B, Lee JH, Bang D. Single-cell RNA sequencing technologies and bioinformatics pipelines. Exp Mol Med 2018; 50(8): 1-14.
[http://dx.doi.org/10.1038/s12276-018-0071-8]

[46]   Gross A, Schoendube J, Zimmermann S, Steeb M, Zengerle R, Koltay P. Technologies for single-cell isolation. Int J Mol Sci 2015; 16(8): 16897-919.
[http://dx.doi.org/10.3390/ijms160816897]

[47]   Jung JY, Lee SH, Kim JM, *et al.* Metagenomic analysis of kimchi, a traditional Korean fermented food. Appl Environ Microbiol 2011; 77(7): 2264-74.
[http://dx.doi.org/10.1128/AEM.02157-10] [PMID: 21317261]

[48]   Ercolini D. Exciting strain-level resolution studies of the food microbiome. Microb Biotechnol 2017; 10(1): 54-6.
[http://dx.doi.org/10.1111/1751-7915.12593] [PMID: 28044418]

[49]   Yin S, Zhu H, Shen M, *et al.* Surface display of heterologous β-galactosidase in food-grade recombinant lactococcus lactis. Curr Microbiol 2018; 75(10): 1362-71.
[http://dx.doi.org/10.1007/s00284-018-1531-z] [PMID: 29922971]

[50]   Pawlak-Szukalska A, Wanarska M, Popinigis AT, Kur J. A novel cold-active β-d-galactosidase with transglycosylation activity from the Antarctic *Arthrobacter* sp. 32cB – Gene cloning, purification and characterization. Process Biochem 2014; 49(12): 2122-33.
[http://dx.doi.org/10.1016/j.procbio.2014.09.018]

[51]   David S, Stevens H, Van Riel M, Simons G, de Vos WM. Leuconostoc lactis β-galactosidase is encoded by two overlapping genes. J Bacteriol 1992; 174(13): 4475-81.
[http://dx.doi.org/10.1128/jb.174.13.4475-4481.1992] [PMID: 1624440]

[52]   Nguyen TH, Splechtna B, Yamabhai M, Haltrich D, Peterbauer C. Cloning and expression of the β-galactosidase genes from *Lactobacillus reuteri* in *Escherichia coli*. J Biotechnol 2007; 129(4): 581-91.
[http://dx.doi.org/10.1016/j.jbiotec.2007.01.034] [PMID: 17360065]

[53]   He X, Han N, Wang YP. Cloning, purification, and characterization of a heterodimeric β-galactosidase from *Lactobacillus kefiranofaciens* ZW3. J Microbiol Biotechnol 2016; 26(1): 20-7.
[http://dx.doi.org/10.4014/jmb.1507.07013] [PMID: 26428731]

[54]   Pfeiler EA, Klaenhammer TR. The genomics of lactic acid bacteria. Trends Microbiol 2007; 15(12): 546-53.
[http://dx.doi.org/10.1016/j.tim.2007.09.010] [PMID: 18024129]

[55]   Perez RH, Zendo T, Sonomoto K. Novel bacteriocins from lactic acid bacteria (LAB): Various structures and applications. Microb Cell Fact 2014; 13(S1) (1): S3.
[http://dx.doi.org/10.1186/1475-2859-13-S1-S3] [PMID: 25186038]

[56]   Camargo AC, Todorov SD, Chihib NE, Drider D, Nero LA. Lactic acid bacteria (LAB) and their bacteriocins as alternative biotechnological tools to control *Listeria monocytogenes* biofilms in food processing facilities. Mol Biotechnol 2018; 609: 712-26.

[57]   Ennahar S, Sashihara T, Sonomoto K, Ishizaki A. Class IIa bacteriocins: Biosynthesis, structure and activity. FEMS Microbiol Rev 2000; 24(1): 85-106.
[http://dx.doi.org/10.1111/j.1574-6976.2000.tb00534.x] [PMID: 10640600]

[58]   Todorov SD. Bacteriocins from *Lactobacillus plantarum* production, genetic organization and mode of action: Produção, organização genética e modo de ação. Braz J Microbiol 2009; 40(2): 209-21.
[http://dx.doi.org/10.1590/S1517-83822009000200001] [PMID: 24031346]

[59]   Liong MT, Shah NP. Bile salt deconjugation ability, bile salt hydrolase activity and cholesterol co-precipitation ability of *Lactobacilli* strains. Int Dairy J 2005; 15(4): 391-8.
[http://dx.doi.org/10.1016/j.idairyj.2004.08.007]

[60]   Gil-Rodríguez AM, Beresford T. Bile salt hydrolase and lipase inhibitory activity in reconstituted skim

milk fermented with lactic acid bacteria. J Funct Foods 2021; 77: 104342.
[http://dx.doi.org/10.1016/j.jff.2020.104342]

[61]   Prete R, Long SL, Gallardo AL, Gahan CG, Corsetti A, Joyce SA. Beneficial bile acid metabolism from *Lactobacillus plantarum* of food origin. Sci Rep 2020; 10(1): 1165.
[http://dx.doi.org/10.1038/s41598-020-58069-5]

[62]   Jones BV, Begley M, Hill C, Gahan CGM, Marchesi JR. Functional and comparative metagenomic analysis of bile salt hydrolase activity in the human gut microbiome. Proc Natl Acad Sci 2008; 105(36): 13580-5.
[http://dx.doi.org/10.1073/pnas.0804437105] [PMID: 18757757]

[63]   Song Z, Cai Y, Lao X, *et al.* Taxonomic profiling and populational patterns of bacterial bile salt hydrolase (BSH) genes based on worldwide human gut microbiome. Microbiome 2019; 7(1): 1-16.

[64]   Soumya MP, Nampoothiri KM. An overview of functional genomics and relevance of glycosyltransferases in exopolysaccharide production by lactic acid bacteria. Int J Biol Macromol 2021; 184: 1014-25.
[http://dx.doi.org/10.1016/j.ijbiomac.2021.06.131] [PMID: 34171260]

[65]   Vinothkanna A, Sathiyanarayanan G, Rai AK, *et al.* Exopolysaccharide produced by probiotic *Bacillus albus* dm-15 isolated from ayurvedic fermented *Dasamoolarishta*: characterization, antioxidant, and anticancer activities. Front Microbiol 2022; 13: 832109.
[http://dx.doi.org/10.3389/fmicb.2022.832109] [PMID: 35308379]

[66]   Zeidan AA, Poulsen VK, Janzen T, *et al.* Polysaccharide production by lactic acid bacteria: From genes to industrial applications. FEMS Microbiol Rev 2017; 41(1) (1): S168-200.
[http://dx.doi.org/10.1093/femsre/fux017] [PMID: 28830087]

[67]   Kumar R, Bansal P, Singh J, Dhanda S. Purification, partial structural characterization and health benefits of exopolysaccharides from potential probiotic *Pediococcus acidilactici* NCDC 252. Process Biochem 2020; 99: 79-86.
[http://dx.doi.org/10.1016/j.procbio.2020.08.028]

[68]   Prete R, Alam MK, Perpetuini G, Perla C, Pittia P, Corsetti A. Lactic acid bacteria exopolysaccharides producers: A sustainable tool for functional foods. Foods 2021; 10(7): 1653.
[http://dx.doi.org/10.3390/foods10071653]

[69]   Diana M, Quílez J, Rafecas M. Gamma-aminobutyric acid as a bioactive compound in foods: A review. J Funct Foods 2014; 10: 407-20.
[http://dx.doi.org/10.1016/j.jff.2014.07.004]

[70]   Xu N, Wei L, Liu J. Biotechnological advances and perspectives of gamma-aminobutyric acid production. World J Microbiol Biotechnol 2017; 33(3): 64.
[http://dx.doi.org/10.1007/s11274-017-2234-5] [PMID: 28247260]

[71]   Small PLC, Waterman SR. Acid stress, anaerobiosis and gadCB: Lessons from *Lactococcus lactis* and *Escherichia coli*. Trends Microbiol 1998; 6(6): 214-6.
[http://dx.doi.org/10.1016/S0966-842X(98)01285-2] [PMID: 9675796]

[72]   Li H, Cao Y. Lactic acid bacterial cell factories for gamma-aminobutyric acid. Amino Acids 2010; 39(5): 1107-16.
[http://dx.doi.org/10.1007/s00726-010-0582-7]

[73]   Cui Y, Miao K, Niyaphorn S, Qu X. Production of gamma-aminobutyric acid from lactic acid bacteria: A systematic review. Int J Mol Sci 2020; 21(3): 995.
[http://dx.doi.org/10.3390/ijms21030995] [PMID: 32028587]

[74]   Wegmann U, O'Connell-Motherway M, Zomer A, *et al.* Complete genome sequence of the prototype lactic acid bacterium *Lactococcus lactis* subsp. cremoris MG1363. J Bacteriol 2007; 189(8): 3256-70.
[http://dx.doi.org/10.1128/JB.01768-06] [PMID: 17307855]

[75]   Lyu C, Zhao W, Peng C, *et al.* Exploring the contributions of two glutamate decarboxylase isozymes

in *Lactobacillus brevis* to acid resistance and γ-aminobutyric acid production. Microb Cell Fact 2018; 17(1): 180.
[http://dx.doi.org/10.1186/s12934-018-1029-1]

[76]   Gao D, Chang K, Ding G, *et al.* Genomic insights into a robust gamma-aminobutyric acid-producer *Lactobacillus brevis* CD0817. AMB Express 2019; 9(1): 72.
[http://dx.doi.org/10.1186/s13568-019-0799-0]

[77]   Ma D, Lu P, Yan C, *et al.* Structure and mechanism of a glutamate–GABA antiporter. Nature 2012; 483(7391): 632-6.
[http://dx.doi.org/10.1038/nature10917]

[78]   Sharma A, Lee S, Park YS. Molecular typing tools for identifying and characterizing lactic acid bacteria: A review. Food Sci Biotechnol 2020; 29(10): 1301-18.
[http://dx.doi.org/10.1007/s10068-020-00802-x]

[79]   Plavec TV, Berlec A. Engineering of lactic acid bacteria for delivery of therapeutic proteins and peptides. Appl Microbiol Biotechnol 2019; 103(5): 2053-66.
[http://dx.doi.org/10.1007/s00253-019-09628-y] [PMID: 30656391]

[80]   Zuo F, Marcotte H. Advancing mechanistic understanding and bioengineering of probiotic *Lactobacilli* and *Bifidobacteria* by genome editing. Curr Opin Biotechnol 2021; 70: 75-82.
[http://dx.doi.org/10.1016/j.copbio.2020.12.015] [PMID: 33445135]

[81]   Marcotte H, Krogh Andersen K, Lin Y, *et al.* Characterization and complete genome sequences of *L. rhamnosus* DSM 14870 and *L. gasseri* DSM 14869 contained in the EcoVag® probiotic vaginal capsules. Microbiol Res 2017; 205: 88-98.
[http://dx.doi.org/10.1016/j.micres.2017.08.003] [PMID: 28942850]

[82]   Wu J, Xin Y, Kong J, Guo T. Genetic tools for the development of recombinant lactic acid bacteria. Microb Cell Fact 2021; 20(1): 118.
[http://dx.doi.org/10.1186/s12934-021-01607-1]

[83]   Tarahomjoo S. Development of vaccine delivery vehicles based on lactic acid bacteria. Mol Biotechnol 2011; 51(2): 183-99.

[84]   Steidler L, Hans W, Schotte L, *et al.* Treatment of murine colitis by *Lactococcus lactis* secreting interleukin-10. Science 2000; 289(5483): 1352-5.

[85]   Stahl B, Barrangou R. Complete genome sequence of probiotic strain *Lactobacillus acidophilus* La-14. Genome Announc 2013; 1(3): e00376-13.
[http://dx.doi.org/10.1128/genomeA.00376-13] [PMID: 23788546]

[86]   Altermann E, Russell WM, Azcarate-Peril MA, *et al.* Complete genome sequence of the probiotic lactic acid bacterium *Lactobacillus acidophilus* NCFM. Proc Natl Acad Sci 2005; 102(11): 3906-12.
[http://dx.doi.org/10.1073/pnas.0409188102] [PMID: 15671160]

[87]   Makarova K, Slesarev A, Wolf Y, *et al.* Comparative genomics of the lactic acid bacteria. Proc Natl Acad Sci 2006; 103(42): 15611-6.
[http://dx.doi.org/10.1073/pnas.0607117103] [PMID: 17030793]

[88]   Hao P, Zheng H, Yu Y, *et al.* Complete sequencing and pan-genomic analysis of *Lactobacillus delbrueckii* subsp. bulgaricus reveal its genetic basis for industrial yogurt production. PLoS One 2011; 6(1): e15964.
[http://dx.doi.org/10.1371/journal.pone.0015964] [PMID: 21264216]

[89]   Sun Z, Chen X, Wang J, *et al.* Complete genome sequence of *Lactobacillus delbrueckii* subsp. *bulgaricus* Strain ND02. J Bacteriol 2011; 193(13): 3426-7.
[http://dx.doi.org/10.1128/JB.05004-11] [PMID: 21515763]

[90]   Morita H, Toh H, Fukuda S, *et al.* Comparative genome analysis of *Lactobacillus reuteri* and *Lactobacillus fermentum* reveal a genomic island for reuterin and cobalamin production. DNA Res 2008; 15(3): 151-61.

[http://dx.doi.org/10.1093/dnares/dsn009] [PMID: 18487258]

[91]   Makarova KS, Koonin EV. Evolutionary genomics of lactic acid bacteria. J Bacteriol 2007; 189(4): 1199-208.
       [http://dx.doi.org/10.1128/JB.01351-06] [PMID: 17085562]

[92]   Callanan M, Kaleta P, O'Callaghan J, *et al.* Genome sequence of *Lactobacillus helveticus*, an organism distinguished by selective gene loss and insertion sequence element expansion. J Bacteriol 2008; 190(2): 727-35.
       [http://dx.doi.org/10.1128/JB.01295-07] [PMID: 17993529]

[93]   Pridmore RD, Berger B, Desiere F, *et al.* The genome sequence of the probiotic intestinal bacterium *Lactobacillus johnsonii* NCC 533. Proc Natl Acad Sci 2004; 101(8): 2512-7.
       [http://dx.doi.org/10.1073/pnas.0307327101] [PMID: 14983040]

[94]   Kleerebezem M, Boekhorst J, Van Kranenburg R, *et al.* Complete genome sequence of *Lactobacillus plantarum* WCFS1. Proc Natl Acad Sci 2003; 100(4): 1990-5.
       [http://dx.doi.org/10.1073/pnas.0337704100] [PMID: 12566566]

[95]   Kankainen M, Paulin L, Tynkkynen S, *et al.* Comparative genomic analysis of *Lactobacillus rhamnosus* GG reveals pili containing a human- mucus binding protein. Proc Natl Acad Sci 2009; 106(40): 17193-8.
       [http://dx.doi.org/10.1073/pnas.0908876106] [PMID: 19805152]

[96]   Chaillou S, Champomier-Vergès MC, Cornet M, *et al.* The complete genome sequence of the meat-borne lactic acid bacterium *Lactobacillus sakei* 23K. Nat Biotechnol 2005; 23(12): 1527-33.

[97]   Claesson MJ, Li Y, Leahy S, *et al.* Multireplicon genome architecture of *Lactobacillus salivarius*. Proc Natl Acad Sci 2006; 103(17): 6718-23.
       [http://dx.doi.org/10.1073/pnas.0511060103] [PMID: 16617113]

[98]   Oh HM, Cho YJ, Kim BK, *et al.* Complete genome sequence analysis of *Leuconostoc kimchii* IMSNU 11154. J Bacteriol 2010; 192(14): 3844-5.
       [http://dx.doi.org/10.1128/JB.00508-10] [PMID: 20494991]

[99]   Bolotin A, Quinquis B, Renault P, *et al.* Complete sequence and comparative genome analysis of the dairy bacterium *Streptococcus thermophilus*. Nat Biotechnol 2004; 22(12): 1554-8.
       [http://dx.doi.org/10.1038/nbt1034]

[100] Kang X, Ling N, Sun G, Zhou Q, Zhang L, Sheng Q. Complete genome sequence of *Streptococcus thermophilus* strain MN-ZLW-002. J Bacteriol 2012; 194(16): 4428-9.
       [http://dx.doi.org/10.1128/JB.00740-12] [PMID: 22843572]

[101] Schwab C, Sørensen KI, Gänzle MG. Heterologous expression of glycoside hydrolase family 2 and 42 β-galactosidases of lactic acid bacteria in *Lactococcus lactis*. Syst Appl Microbiol 2010; 33(6): 300-7.
       [http://dx.doi.org/10.1016/j.syapm.2010.07.002] [PMID: 20822875]

[102] Gangiredla J, Barnaba TJ, Mammel MK, *et al.* Fifty-six draft genome sequences of 10 *Lactobacillus* species from 22 commercial dietary supplements. Genome Announc 2018; 6(26): e00621-18.
       [http://dx.doi.org/10.1128/genomeA.00621-18] [PMID: 29954914]

[103] Poyet M, Groussin M, Gibbons SM, *et al.* A library of human gut bacterial isolates paired with longitudinal multiomics data enables mechanistic microbiome research. Nat Med 2019; 25(9): 1442-52.
       [http://dx.doi.org/10.1038/s41591-019-0559-3]

[104] Piard JC, Muriana PM, Desmazeaud MJ, Klaenhammer TR. Purification and partial characterization of lacticin 481, a lanthionine- containing bacteriocin produced by *Lactococcus lactis* subsp. *lactis* CNRZ 481. Appl Environ Microbiol 1992; 58(1): 279-84.
       [http://dx.doi.org/10.1128/aem.58.1.279-284.1992] [PMID: 16348628]

[105] Hastings JW, Sailer M, Johnson K, Roy KL, Vederas JC, Stiles ME. Characterization of leucocin A-UAL 187 and cloning of the bacteriocin gene from *Leuconostoc gelidum*. J Bacteriol 1991; 173(23):

7491-500.
[http://dx.doi.org/10.1128/jb.173.23.7491-7500.1991] [PMID: 1840587]

[106] Park SH, Itoh K, Kikuchi E, Niwa H, Fujisawa T. Identification and characteristics of nisin Z-Producing *Lactococcus lactis* subsp. *lactis* Isolated from Kimchi. Curr Microbiol 2003; 46(5): 0385-58.

[107] Urso R, Rantsiou K, Cantoni C, Comi G, Cocolin L. Sequencing and expression analysis of the sakacin P bacteriocin produced by a *Lactobacillus sakei* strain isolated from naturally fermented sausages. Appl Microbiol Biotechnol 2006; 71(4): 480-5.
[http://dx.doi.org/10.1007/s00253-005-0172-x]

[108] Zhang X, Shang N, Zhang X, Gui M, Li P. Role of plnB gene in the regulation of bacteriocin production in *Lactobacillus paraplantarum* L-XM1. Microbiol Res 2013; 168(5): 305-10.
[http://dx.doi.org/10.1016/j.micres.2012.11.008] [PMID: 23276644]

[109] McAuliffe O, Cano RJ, Klaenhammer TR. Genetic analysis of two bile salt hydrolase activities in *Lactobacillus acidophilus* NCFM. Appl Environ Microbiol 2005; 71(8): 4925-9.
[http://dx.doi.org/10.1128/AEM.71.8.4925-4929.2005] [PMID: 16085898]

[110] Fang F, Li Y, Bumann M, *et al.* Allelic variation of bile salt hydrolase genes in *Lactobacillus salivarius* does not determine bile resistance levels. J Bacteriol 2009; 191(18): 5743-57.
[http://dx.doi.org/10.1128/JB.00506-09] [PMID: 19592587]

[111] Van der Meulen R, Grosu-Tudor S, Mozzi F, *et al.* Screening of lactic acid bacteria isolates from dairy and cereal products for exopolysaccharide production and genes involved. Int J Food Microbiol 2007; 118(3): 250-8.
[http://dx.doi.org/10.1016/j.ijfoodmicro.2007.07.014] [PMID: 17716765]

[112] Germond JE, Delley M, D'Amico N, Vincent SJF. Heterologous expression and characterization of the exopolysaccharide from *Streptococcus thermophilus* Sfi39. Eur J Biochem 2001; 268(19): 5149-56.
[http://dx.doi.org/10.1046/j.0014-2956.2001.02450.x] [PMID: 11589707]

[113] Knoshaug EP, Ahlgren JA, Trempy JE. Exopolysaccharide expression in *Lactococcus lactis* subsp. *cremoris* Ropy352: Evidence for novel gene organization. Appl Environ Microbiol 2007; 73(3): 897-905.
[http://dx.doi.org/10.1128/AEM.01945-06] [PMID: 17122391]

[114] Sanders JW, Leenhouts K, Burghoorn J, Brands JR, Venema G, Kok J. A chloride-inducible acid resistance mechanism in *Lactococcus lactis* and its regulation. Mol Microbiol 1998; 27(2): 299-310.
[http://dx.doi.org/10.1046/j.1365-2958.1998.00676.x] [PMID: 9484886]

[115] Nomura M, Nakajima I, Fujita Y, *et al. Lactococcus lactis* contains only one glutamate decarboxylase gene. Microbiology 1999; 145(6): 1375-80.
[http://dx.doi.org/10.1099/13500872-145-6-1375] [PMID: 10411264]

[116] Siezen RJ, Starrenburg MJC, Boekhorst J, Renckens B, Molenaar D, Van Hylckama Vlieg JET. Genome-scale genotype-phenotype matching of two *Lactococcus lactis* isolates from plants identifies mechanisms of adaptation to the plant niche. Appl Environ Microbiol 2008; 74(2): 424-36.
[http://dx.doi.org/10.1128/AEM.01850-07] [PMID: 18039825]

[117] Surachat K, Deachamag P, Kantachote D, Wonglapsuwan M, Jeenkeawpiam K, Chukamnerd A. *In silico* comparative genomics analysis of *Lactiplantibacillus plantarum* DW12, a potential gamma-aminobutyric acid (GABA)-producing strain. Microbiol Res 2021; 251: 126833.
[http://dx.doi.org/10.1016/j.micres.2021.126833] [PMID: 34352473]

[118] Kelly WJ, Altermann E, Lambie SC, Leahy SC. Interaction between the genomes of *Lactococcus lactis* and phages of the P335 species. Front Microbiol 2013; 4: 257.
[http://dx.doi.org/10.3389/fmicb.2013.00257] [PMID: 24009606]

[119]   Kato H, Shiwa Y, Oshima K, *et al.* Complete genome sequence of *Lactococcus lactis* IO-1, a lactic acid bacterium that utilizes xylose and produces high levels of L-lactic acid. J Bacteriol 2012; 194(8): 2102-3.
[http://dx.doi.org/10.1128/JB.00074-12] [PMID: 22461545]

CHAPTER 7

L. Plantarum of Vegetable Origin - Genome Editing and Applications

Sudeepa E. S.[1,*] and **A. Sajna**[1]

[1] *Department of Biotechnology, Nehru Arts and Science College, Nehru Gardens, T M Palayam, Coimbatore, Tamil Nadu, India*

Abstract: *Lactobacillus plantarum* is a widespread, versatile bacterium that plays a vital role in the preservation of innumerable fermented foods. These strains are commonly employed as silage additives and starter cultures of fermented goods. Genome editing could provide an added benefit by improving the fermentation profile and quality, as well as the accompanying therapeutic benefits.

Genome editing of various strains of *L. plantarum* can be used commercially to produce L-ribulose or succinic acid, direct lactic acid production, and increased ethanol production. *L. plantarum* strains or recombinant strains can help restore intestinal flora homeostasis, reduce the number of pathogenic organisms, and could even be employed as vaccine carriers. Food products such as raw and fermented vegetables, olives, and cereals inoculated with probiotic microbes have shown encouraging benefits as people now seek non-dairy based probiotics. *L. plantarum* of vegetable or plant origin, as well as applications of genome edited strains, are discussed in this book chapter.

Keywords: Fermentation, Genome rditing, *Lactiplantibacillus plantarum*, *Lactobacilli*, Probiotics.

INTRODUCTION

Lactic acid bacteria are one of the most significant food-grade microorganisms, with nutritional and fermentative properties [1]. *L. plantarum* is a member of the *Lactobacillus casei – Pediococcus* phylogenetic group [2], specifically the *L. plantarum* taxonomic subgroup. Many metabolic processes distinguish this species, allowing it to colonise a variety of environments, including dairy products, pickled vegetables, fish products, silage, wine, and mammalian intestinal tracts [3].

Corresponding author Sudeepa E. S.: Department of Biotechnology, Nehru Arts and Science College, Nehru Gardens, T M Palayam, Coimbatore, Tamil Nadu, India; E-mail: sudeepa.es@gmail.com

Prakash M. Halami & Aravind Sundararaman (Eds.)

L. plantarum strains are utilised as a starting culture in sourdough bread, meat products, and wine because they are associated with favourable qualities in many fermented foods [4] and are added to a number of them to improve their quality or associated health benefits, as a result, paving way for the availability of a variety of bio-therapeutic products [5]. They play an essential role in the preservation of food and fermented food products such as vegetables, sausages, silages, brine olives, sauerkraut, cassava, and kimchi [6]. Due to its higher acid tolerance than other lactic acid bacteria, it can significantly improve the latter stages of fruit and vegetable fermentations [7]. As a result, these bacteria are frequently used in industrial fermentation and are "generally recognised as safe" (GRAS), which qualifies the safety presupposition [8].

Various functional features of microorganisms depend on its genes. Dissecting biochemical processes that drive food fermentation, as well as identifying and characterisation of health-promoting traits that have a favourable impact on the composition and roles of microbiomes in human health, are all dependent on the genome of the microbe. Our ability to manipulate genomes, given the recent advancements in molecular biology, has made easier to tinker with the genes and gain the desired modifications.

Genome editing precisely defined as a genetic engineering principle which deals with the alterations of genome of an organism at single base pair levels in order to manipulate or add on or enhance particular features of the organism [9]. Genome editing has emerged as a widely practised engineering tool in various microorganisms like *Escherichia coli, Staphylococcus aureus, Lactobacillus reuteri, Clostridium beijerinckii, Streptococcus pneumonia,* and *Saccharomyces cerevisiae etc.* as it is a highly efficient and simple tool for site specific manipulation of genes when compared to Zinc Finger Nucleases (ZFNs) and Transcription Activator-Like Effector Nucleases (TALENs) techniques [10].

CRISPR-Cas9 genome-editing mechanism has evolved naturally in prokaryotes like bacteria as a protective mechanism against bacteriophages. CRISPR arrays are DNA segments created by bacteria that catch fragments of DNA from invading viruses [11]. CRISPR-Cas9 assisted double-stranded DNA and single-stranded DNA recombination engineering in *L. plantarum* WCFS1 was used to edit the genome, including gene knockouts, insertions, and point mutations [12]. Most of the *L. plantarum* have the type II CRISPR-Cas system, comprising four genes- Cas1, Cas2, Cas9, and csn2 [13].

With the development of clustered, regularly interspaced short palindromic repeat (CRISPR) based technologies, we are on the verge of a broad-scale genome editing revolution. Various genome editing tools and technologies, such as

CRISPR-associated enzyme genome editing, single-stranded DNA recombinant engineering, and bacteriophage modification, have paved the way for the development of next-generation bio therapeutic agents with improved genotypes and health-promoting functional features.

Genetic manipulation of *Lactobacillus sp.* could be beneficial to get an improved fermentation profile and or enhance medically important characters, *etc* [14]. The main aim of the book chapter is to put emphasis on the various applications of genome editing in *L. plantarum.*

LACTIPLANTIBACILLUS PLANTARUM

Morphology

L. plantarum is a nomadic, non-pathogenic gram-positive lactic acid bacterium previously named *Streptobacterium plantarum* [15]. *L. plantarum* is named after the Latin word *"plantarum"* which means "plant species." They are non-motile, non-sporing forming, catalase negative, rod-shaped and appear individually or in pairs, or in short chains. When haem is present, certain strains exhibit pseudo-catalase activity or true catalase activity.

The scientific classification of *L. plantarum* is given as follows (Basonym: *Lactobacillus plantarum*) in Table **1**. It has two main subspecies associated with it - *L. plantarum* (subsp. *plantarum* and subsp. *argentoratensis*) [16].

Table 1. Classification of *L. plantarum* [18].

Domain	Bacteria
Phylum	*Firmicutes*
Class	*Bacilli*
Order	*Lactobacillales*
Family	*Lactobacillaceae*
Genus	*Lactiplantibacillus*
Species	*L. plantarum*
Subspecies	*L. plantarum* subsp. *Argentoratensis* *L. plantarum* subsp. *Plantarum*

L. plantarum is an aero-tolerant *bacteriium* that grows well at 15°C [17] in a 4% NaCl concentration but not much at 45°C, according to research. It can convert hexoses into both D and L isomers of lactic acid *via* fermentation, as well as pentoses and/or gluconate, into acetic and lactic acids. It also has the ability to ferment malic acid to lactic acid and carbon dioxide, as well as citric acid to diacetyl, acetoin, and carbon dioxide. These bacteria can transition between hetero-fermentative and homo-fermentative metabolism depending on the carbon supply.

L. plantarum is mostly thought to be a homo-fermentative bacteria [19, 20]. It has a strong tolerance for low pH, demonstrating its acid resilience [21]. As a result, it predominates spontaneously in lactic acid fermented foods and other plant-based foods, such as brined olives [22], capers [23], sauerkraut, salted gherkins [24], sourdough, Nigerian ogi, Ethiopian sourdough made from the cassava [24, 25] *etc.* as cited in Table **2**. The pH of these foods is typically below 4.0. It can also withstand the acidic environment of the human stomach and is considered bile tolerant [26]. This indicates that individuals consuming such plant-origin fermented food items are also consuming considerable amounts of *L. plantarum* one way or the other.

Table 2. Sources of various species/strains of *L. plantarum* [29].

Strain	Source of Isolation	Origin/Native
L. plantarum subsp. *plantarum*		
B41	Silage	Italy
CNRZ 1220	Cheese	Egypt
LAB R1M3	Monte Veronese cheese	Italy
ZW20	Cheese	Switzerland
L. plantarum subsp. *argentoratensis*		
A, 57.2, A4, A7	Cassava	Colombia
DK 19	White maize	Nigeria
NCIMB 12120, CNRZ 1889	Fermented cereals (ogi)	Nigeria
CNRZ 1890	Fermented millet (baba)	Nigeria
DK 9	Fermented cucumber	Nigeria
DK 36	Tapioca	Nigeria
LP85-2	Silage	France
SF2A35B	Sour cassava	South America
L. paraplantarum		
LMG 16673T	Beer, France	France

Researchers and the food sector are of current interest in *L. plantarum* because it is a safe probiotic and has the potential to reduce the number of pathogens or diseases that can affect humans. Furthermore, the current study suggests that it could also be employed as a vaccine carrier.

Consumer demand for healthy, fresh, nutritious, and minimally processed food is currently very high. Various techniques are used to maintain the required quality or characteristics of food products, as well as to retain or extend their shelf-life, the simplest of which is fermentation. Recent developments show a shift in the quest for non-dairy fermented food products for probiotic systems, particularly based on fruits, vegetables, and cereals [27]. The balance of acetic (volatile) and lactic (non-volatile) acids by the bacterium gives the food a specific taste and flavour. Although the bacterium is mostly employed for dough acidification, it can also help with the process of leavening.

Sauerkraut is an example of a food item made by *L. plantarum* fermentation. Sauerkraut is made naturally from cabbage by lactic acid bacteria [28], which are acid tolerant homolactic fermenters. Sauerkraut fermented with *L. plantarum*, on the other hand, has an acidic, vinegar like flavour. As a result, these bacteria improved the final sauerkraut product by reducing fermentation time and harmful microbe development.

Genome

L. plantarum has one of the biggest genomes among lactic acid bacteria, owing to its versatility. In 2003, the strain WCFS1, which was isolated from human saliva, became the first *L. plantarum* strain to be completely sequenced. A circular chromosome of 3.3 Mb makes up its genome, as shown in Fig. (**1**). There are five rRNA operons on the chromosome, which are uniformly dispersed. A total of 62 tRNA encoding genes have been discovered, some of which are linked to rRNA clusters. Furthermore, the genome contains two types of transposase sites, both of which are assumed to encode mobile genetic elements. Only 39 of the 3,052 protein-encoding genes found in the genome are pseudogenes. The strain ATCC14917T is the first strain isolated from fermented foods, sequenced in 2009. Since then, a number of fresh *L. plantarum* genomes isolated from various sources have been sequenced and are available in biological databases.

Metabolism

L. plantarum contain low GC content, a peptidoglycan cell wall, and are arranged collinearly. *L. plantarum* proteins are extremely similar to those of other Gram-positive bacteria. The genes coding for the Embden-Meyerhoff-Parnas (EMP) pathway, as shown in Fig. (**2**), are found in the *L. plantarum* genome. Enzymes that break down pentoses and hexoses are correlated to some of the genes. Furthermore, its genome reveals genes that code for the phosphotransferase, mannose, and fructose transport mechanisms [31].

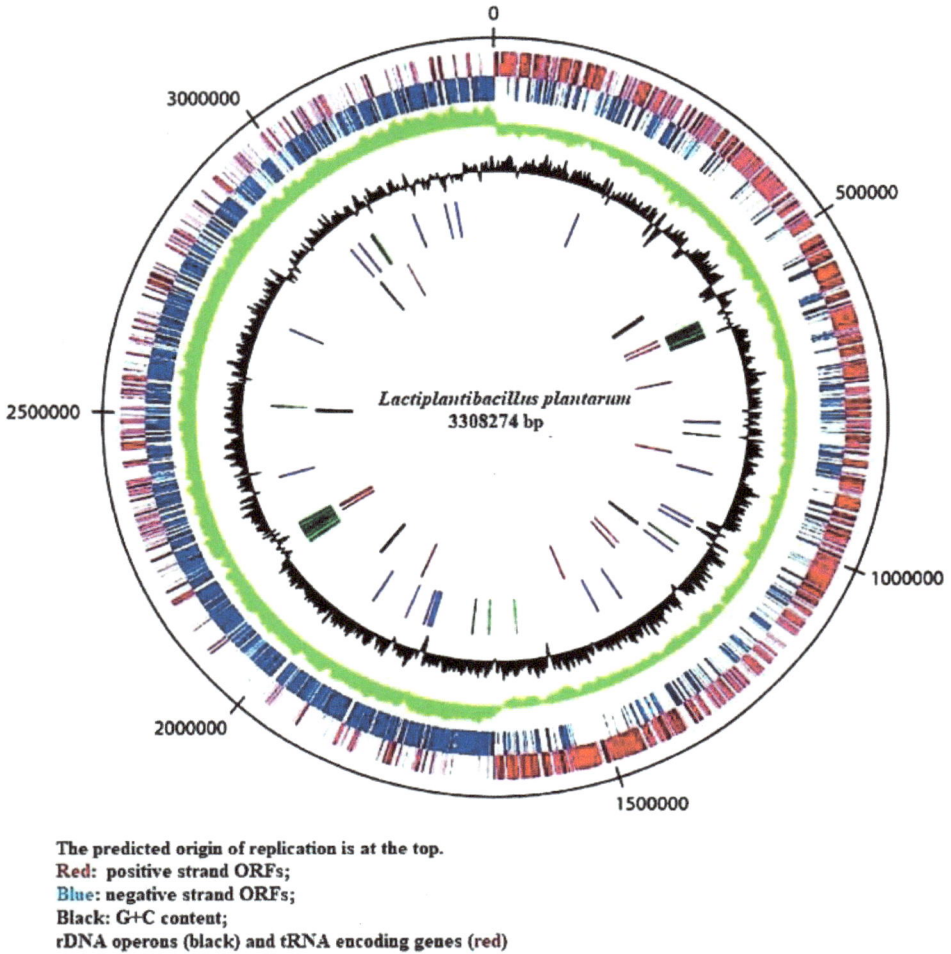

Fig. (1). *L. plantarum* strain WCFS1 genome (3.3 kb) isolated from human saliva [30].

L. PLANTARUM FERMENTED VEGETABLE PRODUCTS

Due to its capacity to withstand the high salt and acidic content of fermented vegetables, such as cucumber, sauerkraut, and olive [29], as depicted in Table **2**, *L. plantarum* is the most prevalent species or strains detected in fermented vegetables. Aside from that, *L. plantarum* strains are used as starter cultures for a variety of fermented vegetables.

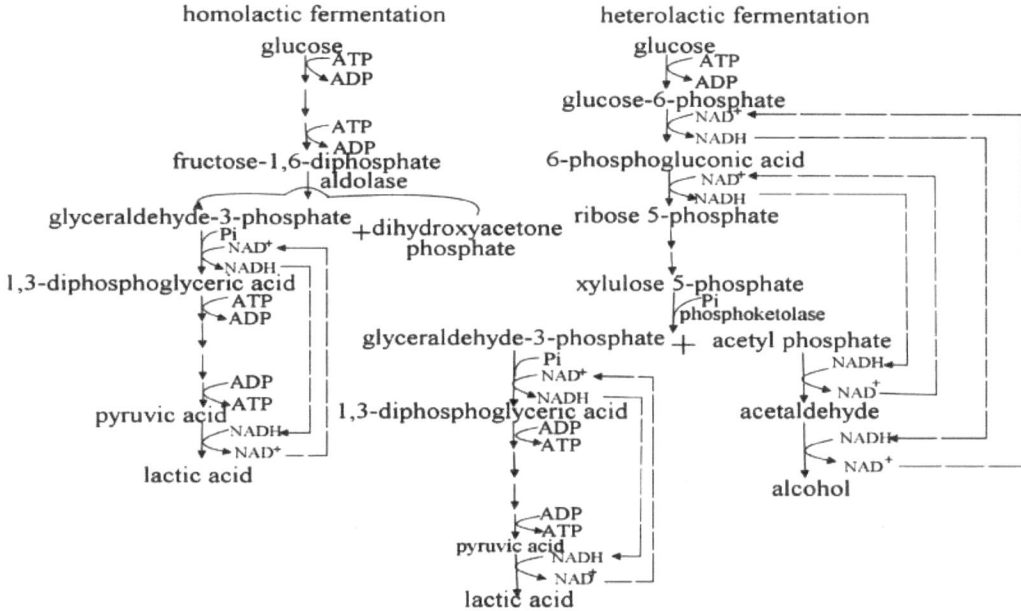

Fig. (2). Metabolic pathways in *L. plantarum* [31].

Several studies have been published on fermented sweet potato (*Ipomoea batatas*) food products using *L. plantarum* (MTCC 1407) as the starter culture and the consumers admired the pickled sweet potato that was developed [32, 33]. In addition, *L. plantarum* B282 strains have been effectively used as starters in the Spanish style fermentation of green olive [34]. Also, *L. plantarum* was the first lactic acid bacteria to be found associated with cucumber fermentation [35].

Kimchi is a fermented Korean food made from Chinese cabbage. According to its properties as compared to commercial probiotics and yogurt starter strains, *L. plantarum* Ln4, one of the numerous strains isolated from kimchi, could be used as a probiotic [36]. A probiotic trait was found in *L. plantarum* wikim 18 (KFCC 11588P) isolated from 'baechu' (napa cabbage). The effect of *L. plantarum* ATCC 8014 fermentation on the oxidative stability of olive oil has been proposed as a suitable approach for preserving olive quality and stability during storage [37]. *L. plantarum* S0/7 was also found to have probiotic characteristics [38].

Certain researches show the isolation of *L. plantarum* from starch rich cassava wastes. 'Fufu' is a fermented wet paste made from cassava starch that is widely consumed in West Africa. Because there are some drawbacks to spontaneous food fermentation, starter cultures are preferred. The use of *L. plantarum* (strain 6710) as a starter in the fermentation of fortified cassava flour with protein and pro-vitamin A, resulted in the formation of wet fufu, which was positively received by customers [39]. Scientists have recently optimised the process parameters, which include inoculum amount, salt concentration, and incubation period for pickling elephant foot yam (*Amorphophallus paeoniifolius*).

PROPERTIES OF PLANT-BASED FOOD PRODUCTS FERMEMTED BY *L. PLANTARUM*

Taste/Flavour Attributes

The quality of fermented food products (*e.g.*, *kimchi*, *sauerkraut*, *jeotgal*, and *pickles*) could be improved by several *L. plantarum* strains, in terms of stability, quality, enhanced taste, and health-promoting benefits.

Exopolysaccharides Production

EPS are "food grade biopolymers" or high molecular weight extracellular biopolymers derived from natural sources and produced by microorganisms like bacteria, fungi, and blue-green algae during their metabolic processes [40]. Lactic acid bacteria are often regarded as significant EPS-producing microbes that are also safe to eat [41]. Furthermore, EPS has been shown to contribute to the viscoelastic of fermented foods and to have potential health-promoting effects in the development of functional meals.

L. plantarum is a well-known microbe for its potential to produce EPS, and it has received a huge amount of attention. The EPS-producing bacteria have a big influence on the monosaccharide content and structure, as well as the microbial culture conditions and media composition. EPS produced by lactic acid bacteria has been identified as a potential grade of bioactive natural compounds in biochemical and medicinal applications, as they have immuno-modulatory [42], anticancer, and antioxidant properties, as well as cholesterol-lowering capabilities. *L. plantarum* ZDY2013 EPS might be a viable medicinal and health food candidate.

Bacteriocin Production

Bacteriocins are a group of genetically programmed antibacterial peptides that are active against both related and pathogenic as well as food spoilage microorganisms [43]. *L. plantarum* generates bacteriocin with high activity and antibiotic activity against *S. aureus*, *L. monocytogenes,* and *A. hydrophila,* among other bacteria [44].

L. plantarum BFE 905 developed a bacteriocin called 'plantaricin D' that was potent against *Lactobacillus sakei* and *Listeria monocytogenes* strains [45]. Plantaricin 163 was also discovered to be generated by *L. plantarum* 163, which was isolated from traditional Chinese fermented vegetables [46]. MBSa4, a plantaricin generated by *L. plantarum* with a molecular weight of 2.5 kDa, was identified from Brazilian salami [47]. Similarly, a novel bacteriocin-M1-UVs300, which was produced by *L. plantarum* M1-UVs300, was purified and characterized from fermented sausage [48]. Gram-positive and Gram-negative bacteria were both inhibited by this bacteriocin. Another bacteriocin generated by *L. plantarum* (NTU 102) was discovered in homemade Korean cabbage pickles. This strain had good acidic (low pH) survival, high bile vigour, increased tolerance/resistance to *Vibrio alginolyticus* infection, pathogen restriction, and a good ability to reduce low-density lipoprotein cholesterol (LDL-C) to high-density lipoprotein cholesterol (HDL-C) ratios.

Bacteriocins produced by *L. plantarum* have a variety of possible applications, including bio-preservation in the food industry, application of crude bacteriocin directly to food products, and utilisation of a previously fermented product from a bacteriocin generating strain.

Probiotic and Health Benefits

When administered in adequate proportions, probiotics are viable microorganisms that provide health benefits/aids to the host. Because of their ability to delay or block the growth of common contaminating microbes, probiotic bacteria can help food products last longer and safer. Increasing health awareness regarding consumption of probiotic strains has been encouraged among consumers to overcome the growing diseases risks like intolerance to lactose, diabetes, and cardiovascular diseases.

The scientific validity of *L. plantarum* strain as probiotics was first determined by evaluating bile and acid tolerance in animal and human digestive tracts. Furthermore, it showed a significant reduction in infection and GI tract diseases [49], such as Crohn's disease, inflammatory bowel disease (IBD), and colitis [50, 51]. This bacterium is thought to have favourable health effects as a probiotic, particularly in terms of inhibiting pathogen attachment to intestinal cell lines. Its practicable application includes the addition of a variety of fermented foods and beverages. When the probiotic capabilities of a variety of *L. plantarum* strains of vegetable origin were investigated, it was discovered that probiotic properties varied between strains [52]. Probiotic strains of *L. plantarum* have been shown to exhibit therapeutic properties and are being used to treat a variety of chronic diseases, including Alzheimer's, Parkinson's, diabetes, obesity, cancer, hypertension, urogenital disorders, and liver ailments [53]. For example, kimchi is well-known as a healthy dish that has health-promoting nature such as anticancer, anti-oxidative, anti-diabetic, and anti-obesity properties.

APPLICATIONS OF GENOME EDITING IN *L. PLANTARUM*

A genome edited *L. plantarum* WCFS1 can be used to produce high quantities of N-acetyl glucosamine and enhance its production [12]. A genome editing procedure like this could help expand the methods for accurate and diversified genetic engineering in lactic acid bacteria, allowing strain engineering to be used for more applications.

Food grade recombinants were generated by inserting an exogenous gene *nisI* into the chromosome of *L. plantarum* Lp80 and *L. plantarum* LZ, improving their ability to survive through fermentation operation. Because this bacterium is utilised as a starter, the gene would express nisin immunity, which could be a beneficial defence against naturally occurring nisin makers in the fermenting substrates as well as additional nisin preservatives. Rossi *et al.* [54] discovered that the genetically edited variants were able to exhibit nisin immunity at a higher level than the wild type and did not differ in growth capacity from the corresponding *L. plantarum* parental strains. For the intrinsic stability of the mutation produced and the simplicity of the cloning technique, vector-free chromosomal integration is preferred among other food-grade cloning and genome editing strategies.

Using genetically edited/altered *L. plantarum* to produce porcine lactoferrin, Xu *et al.* [55] showed encouraging results in providing a beneficial agent for weaned piglets against weaning stress and infection. This porcine lactoferrin, which was produced and utilised as a feed additive in the daily diet of weaned piglets, raised the average daily feed intake, enhanced feed efficiency, decreased infection and diarrhoea incidence, and increased the range of vital gut microbiota.

Apart from being employed as probiotics, it was discovered that specific GM strains of *L. plantarum* that can generate and release SOD have anti-inflammatory benefits in a TNBS-induced colitis model [56].

The use of genetically edited *L. plantarum* for the fermentation of lignocellulosic biomass such as corn stover and sorghum stalks increases D-lactic acid yield, demonstrating the use of renewable lignocellulosic biomass as an alternative to conventional feedstock with genome edited lactic acid bacteria to produce D-lactic acid [57].

Similarly, genome editing and metabolic engineering of *L. plantarum* strains to produce recombinant strains can be used commercially to produce a variety of products such as L-ribulose [58], succinic acid [59], direct lactic acid production [60], and even enhanced ethanol production [61]. As a result, novel genome editing and metabolic engineering tools may help to expand the range of genetic alteration approaches accessible in *L. plantarum* and boost commercial applications.

FUTURE PROSPECTS

To date, the focus has been on creating probiotic systems based solely on dairy products. This trend has recently shifted, and large-scale studies are now being undertaken outside the traditional framework. Fermented dairy products include a significant amount of cholesterol and lactose, which may limit their use in people who are lactose intolerant or allergic to lactose. As a result, additional probiotic strains are being discovered in non-dairy items, particularly vegetables and fruits, because they pose no possible risks to consumers.

Fermented juices and vegetables, minimally processed fruits and vegetables, and others are examples of probiotic foods derived from vegetables. Fermented and non-fermented fruits and vegetables, organic probiotic drinks, and dried fruit packed with these microbes are all commercially available. Microencapsulation, vacuum impregnation, and immobilisation technologies have all been utilised to integrate bioactive molecules like probiotics with promising results [62].

These investigations would show the interaction of probiotics carried by vegetable-derived products. Further research on the use of *L. plantarum* strains as a probiotic in non-dairy fermented food products is needed. Its strains are currently being studied for integration into traditional fermented food systems. More innovations to use *L. plantarum*'s probiotic property to promote human health are likely to be made in the next years. Further, innovative methods in gene editing and remarkable applications of these strains need to be researched.

CONCLUSION

L. plantarum, a potential probiotic bacterium readily available in native fermented foods of vegetable origin, possesses a significant ability to distinguish between diverse harmful bacteria, including gram-negative and gram-positive species that can contaminate food and cause human sickness. It can help in restoring intestinal flora equilibrium, reduce the number of harmful bacteria, and could even be employed as vaccine carriers. They are also used as a biocontrol agent to keep infections at bay. To limit the use of medicaments, this probiotic strain found in fermented foods can help to improve consumer health and well-being while also lowering risk.

The use of various genome editing technologies has further improved or modified the strains to enhance fermentation capacity, promote human health in the form of probiotics, production of biomass and feedstock, manufacturing of various carbohydrate sources, *etc.*, for commercial use.

ACKNOWLEDGEMENTS

We are grateful to Dr. Prakash Halami, Chief Scientist & Professor-AcSIR, CSIR-Central Food Technological Research Institute, Mysore for motivating us write this particular review.

REFERENCES

[1]   Leroy F, De Vuyst L. Lactic acid bacteria as functional starter cultures for the food fermentation industry. Trends Food Sci Technol 2004; 15(2): 67-78.
        [http://dx.doi.org/10.1016/j.tifs.2003.09.004]

[2]   Hammes WP, Vogel RF. The genus Lactobacillus. Genera Lact Acid Bact 1995; 19-54.

[3]   Vescovo M, Orsi C, Scolari G, Torriani S. Inhibitory effect of selected lactic acid bacteria on microflora associated with ready-to-use vegetables. Lett Appl Microbiol 1995; 21(2): 121-5.
        [http://dx.doi.org/10.1111/j.1472-765X.1995.tb01022.x] [PMID: 7639993]

[4]   Yilmaz B, Bangar SP, Echegaray N, Suri S, Tomasevic I, Lorenzo JM, *et al.* The Impacts of *L. plantarum* on the functional properties of fermented foods: A review of current knowledge. Microorganisms 2022; 10(4).
        [http://dx.doi.org/10.3390/microorganisms10040826] [PMID: 35456875]

[5]   Devi SM, Aishwarya S, Halami PM. Discrimination and divergence among *Lactobacillus plantarum*-group (LPG) isolates with reference to their probiotic functionalities from vegetable origin. Syst Appl Microbiol 2016; 39(8): 562-70.
        [http://dx.doi.org/10.1016/j.syapm.2016.09.005] [PMID: 27729171]

[6]   Guidone A, Zotta T, Ross RP, *et al.* Functional properties of *Lactobacillus plantarum* strains: A multivariate screening study. Lebensm Wiss Technol 2014; 56(1): 69-76.
        [http://dx.doi.org/10.1016/j.lwt.2013.10.036]

[7]   Khemariya P, Singh S, Jaiswal N, Chaurasia SNS. Isolation and identification of *Lactobacillus plantarum* from vegetable samples. Food Biotechnol 2016; 30(1): 49-62.
        [http://dx.doi.org/10.1080/08905436.2015.1132428]

[8]   Ricci A, Allende A, Bolton D, *et al.* Update of the list of QPS-recommended biological agents intentionally added to food or feed as notified to EFSA 5: Suitability of taxonomic units notified to EFSA until September 2016. EFSA J 2017; 15(3): e04663.
        [PMID: 32625420]

[9]   Alexander WG. A history of genome editing in *Saccharomyces cerevisiae*. Yeast 2018; 35(5): 355-60.
        [http://dx.doi.org/10.1002/yea.3300] [PMID: 29247562]

[10]   Javed MR, Sadaf M, Ahmed T, *et al.* CRISPR-Cas System: History and prospects as a genome editing tool in microorganisms. Curr Microbiol 2018; 75(12): 1675-83.
        [http://dx.doi.org/10.1007/s00284-018-1547-4] [PMID: 30078067]

[11]   Khanzadi MN, Khan AA. CRISPR/Cas9: Nature's gift to prokaryotes and an auspicious tool in

genome editing. J Basic Microbiol 2020; 60(2): 91-102.
[http://dx.doi.org/10.1002/jobm.201900420] [PMID: 31693214]

[12]   Zhou D, Jiang Z, Pang Q, Zhu Y, Wang Q, Qi Q. CRISPR/Cas9-assisted seamless genome editing in *Lactobacillus plantarum* and its application in N-acetylglucosamine production. Appl Environ Microbiol 2019; 85(21): e01367-19.
[http://dx.doi.org/10.1128/AEM.01367-19] [PMID: 31444197]

[13]   Cammarota M, De Rosa M, Stellavato A, Lamberti M, Marzaioli I, Giuliano M. *In vitro* evaluation of *Lactobacillus plantarum* DSMZ 12028 as a probiotic: Emphasis on innate immunity. Int J Food Microbiol 2009; 135(2): 90-8.
[http://dx.doi.org/10.1016/j.ijfoodmicro.2009.08.022] [PMID: 19748696]

[14]   Evanovich E, De Souza Mendonça Mattos PJ, Guerreiro JF. Comparative genomic analysis of *Lactobacillus plantarum* : An overview. Int J Genomics 2019; 2019: 1-11.
[http://dx.doi.org/10.1155/2019/4973214] [PMID: 31093491]

[15]   Bergey DH. Manual of determinative bacteriology. South Med J 1926; 19(3): 236.
[http://dx.doi.org/10.1097/00007611-192603000-00027]

[16]   Swain MR, Anandharaj M, Ray RC, Parveen Rani R. Fermented fruits and vegetables of Asia: A potential source of probiotics. Biotechnol Res Int 2014; 2014: 1-19.
[http://dx.doi.org/10.1155/2014/250424] [PMID: 25343046]

[17]   Balasubramanian B, Soundharrajan I, Al-Dhabi NA, *et al.* Probiotic characteristics of *Ligilactobacillus salivarius* AS22 isolated from sheep dung and its application in corn-fox tail millet silage. Appl Sci 2021; 11(20): 9447.
[http://dx.doi.org/10.3390/app11209447]

[18]   Zheng J, Wittouck S, Salvetti E, *et al.* A taxonomic note on the genus *Lactobacillus*: Description of 23 novel genera, emended description of the genus *Lactobacillus* Beijerinck 1901, and union of *Lactobacillaceae* and *Leuconostocaceae*. Int J Syst Evol Microbiol 2020; 70(4): 2782-858.
[http://dx.doi.org/10.1099/ijsem.0.004107] [PMID: 32293557]

[19]   Siezen RJ, Van Hylckama Vlieg JET. Genomic diversity and versatility of *Lactobacillus plantarum*, a natural metabolic engineer. Microb Cell Fact 2011; 10(Suppl 1) (1): S3.
[http://dx.doi.org/10.1186/1475-2859-10-S1-S3] [PMID: 21995294]

[20]   Poldermans B. Food biotechnology: Microorganisms. Trends Food Sci Technol 1995; 6(10): 352.
[http://dx.doi.org/10.1016/S0924-2244(00)89188-8]

[21]   Gonzalez MJF, García PG, Fernández AG, Quintana MCD. Microflora of the aerobic preservation of directly brined green olives from *Hojiblanca cultivar*. J Appl Bacteriol 1993; 75(3): 226-33.
[http://dx.doi.org/10.1111/j.1365-2672.1993.tb02770.x] [PMID: 8244900]

[22]   Pérez Pulido R, Ben Omar N, Abriouel H, Lucas López R, Martínez Cañamero M, Gálvez A. Microbiological study of lactic acid fermentation of Caper berries by molecular and culture-dependent methods. Appl Environ Microbiol 2005; 71(12): 7872-9.
[http://dx.doi.org/10.1128/AEM.71.12.7872-7879.2005] [PMID: 16332762]

[23]   McDonald LC, Shieh DH, Fleming HP, McFeeters RF, Thompson RL. Evaluation of malolactic-deficient strains of *Lactobacillus plantarum* for use in cucumber fermentations. Food Microbiol 1993; 10(6): 489-99.
[http://dx.doi.org/10.1006/fmic.1993.1054]

[24]   Moorthy SN, Mathew G. Cassava fermentation and associated changes in physicochemical and functional properties. Crit Rev Food Sci Nutr 1998; 38(2): 73-121.
[http://dx.doi.org/10.1080/10408699891274174] [PMID: 9526681]

[25]   Johansson ML, Molin G, Jeppsson B, Nobaek S, Ahrné S, Bengmark S. Administration of different *Lactobacillus* strains in fermented oatmeal soup: *In vivo* colonization of human intestinal mucosa and effect on the indigenous flora. Appl Environ Microbiol 1993; 59(1): 15-20.

[http://dx.doi.org/10.1128/aem.59.1.15-20.1993] [PMID: 8439146]

[26] Behera SS, Ray RC, Zdolec N. *Lactobacillus plantarum* with functional properties: An approach to increase safety and shelf-life of fermented foods. Biomed Res Int 2018; 2018.

[27] Liu CJ, Wang R, Gong FM, *et al.* Complete genome sequences and comparative genome analysis of *Lactobacillus plantarum* strain 5-2 isolated from fermented soybean. Genomics 2015; 106(6): 404-11.
[http://dx.doi.org/10.1016/j.ygeno.2015.07.007] [PMID: 26212213]

[28] Bringel F, Castioni A, Olukoya DK, Felis GE, Torriani S, Dellaglio F. *Lactobacillus plantarum* subsp. *argentoratensis* subsp. nov., isolated from vegetable matrices. Int J Syst Evol Microbiol 2005; 55(4): 1629-34.
[http://dx.doi.org/10.1099/ijs.0.63333-0] [PMID: 16014493]

[29] Ray RC, Joshi VK. Fermented foods: Past, present and future.Microorg Ferment Tradit Foods. CRC Press 2014; pp. 1-36.

[30] Kleerebezem M, Boekhorst J, Van Kranenburg R, *et al.* Complete genome sequence of *Lactobacillus plantarum* WCFS1. Proc Natl Acad Sci 2003; 100(4): 1990-5.
[http://dx.doi.org/10.1073/pnas.0337704100] [PMID: 12566566]

[31] Wang Y, Wu J, Lv M, *et al.* Metabolism characteristics of lactic acid bacteria and the expanding applications in food industry. Front Bioeng Biotechnol 2021; 9: 612285.
[http://dx.doi.org/10.3389/fbioe.2021.612285] [PMID: 34055755]

[32] Panda SH, Panda S, Sethuraman Sivakumar P, Ray RC. Anthocyanin-rich sweet potato lacto-pickle: Production, nutritional and proximate composition. Int J Food Sci Technol 2009; 44(3): 445-55.
[http://dx.doi.org/10.1111/j.1365-2621.2008.01730.x]

[33] Panda SH, Ray RC. Lactic acid fermentation of β-carotene rich sweet potato (*Ipomoea batatas L.*) into lacto-juice. Plant Foods Hum Nutr 2007; 62(2): 65-70.
[http://dx.doi.org/10.1007/s11130-007-0043-y] [PMID: 17370124]

[34] Blana VA, Polymeneas N, Tassou CC, Panagou EZ. Survival of potential probiotic lactic acid bacteria on fermented green table olives during packaging in polyethylene pouches at 4 and 20 °C. Food Microbiol 2016; 53(Pt B): 71-5.
[http://dx.doi.org/10.1016/j.fm.2015.09.004] [PMID: 26678132]

[35] Pérez-Díaz IM, Hayes J, Medina E, *et al.* Reassessment of the succession of lactic acid bacteria in commercial cucumber fermentations and physiological and genomic features associated with their dominance. Food Microbiol 2017; 63: 217-27.
[http://dx.doi.org/10.1016/j.fm.2016.11.025] [PMID: 28040172]

[36] Son SH, Jeon HL, Jeon EB, *et al.* Potential probiotic *Lactobacillus plantarum* Ln4 from kimchi: Evaluation of β-galactosidase and antioxidant activities. Lebensm Wiss Technol 2017; 85: 181-6.
[http://dx.doi.org/10.1016/j.lwt.2017.07.018]

[37] Hamid abadi Sherahi M, Shahidi F, Yazdi FT, Hashemi SMB. Effect of *Lactobacillus plantarum* on olive and olive oil quality during fermentation process. Lebensm Wiss Technol 2018; 89: 572-80.
[http://dx.doi.org/10.1016/j.lwt.2017.10.025]

[38] Saelim K, Jampaphaeng K, Maneerat S. Functional properties of *Lactobacillus plantarum* S0/7 isolated fermented stinky bean (Sa Taw Dong) and its use as a starter culture. J Funct Foods 2017; 38: 370-7.
[http://dx.doi.org/10.1016/j.jff.2017.09.035]

[39] Rosales-Soto MU, Gray PM, Fellman JK, *et al.* Microbiological and physico-chemical analysis of fermented protein-fortified cassava (*Manihot esculenta* Crantz) flour. Lebensm Wiss Technol 2016; 66: 355-60.
[http://dx.doi.org/10.1016/j.lwt.2015.10.053]

[40] Wang J, Zhao X, Tian Z, Yang Y, Yang Z. Characterization of an exopolysaccharide produced by *Lactobacillus plantarum* YW11 isolated from Tibet Kefir. Carbohydr Polym 2015; 125: 16-25.

[http://dx.doi.org/10.1016/j.carbpol.2015.03.003] [PMID: 25857955]

[41]   Zhang Z, Liu Z, Tao X, Wei H. Characterization and sulfated modification of an exopolysaccharide from *Lactobacillus plantarum* ZDY2013 and its biological activities. Carbohydr Polym 2016; 153: 25-33.
[http://dx.doi.org/10.1016/j.carbpol.2016.07.084] [PMID: 27561468]

[42]   Zhou K, Zeng Y, Yang M, *et al.* Production, purification and structural study of an exopolysaccharide from *Lactobacillus plantarum* BC-25. Carbohydr Polym 2016; 144: 205-14.
[http://dx.doi.org/10.1016/j.carbpol.2016.02.067] [PMID: 27083810]

[43]   Hu Y, Liu X, Shan C, *et al.* Novel bacteriocin produced by *Lactobacillus alimentarius* FM-MM 4 from a traditional Chinese fermented meat Nanx Wudl: Purification, identification and antimicrobial characteristics. Food Control 2017; 77: 290-7.
[http://dx.doi.org/10.1016/j.foodcont.2017.02.007]

[44]   Choi S, Gyung BM, Chung MJ, Lim S, Yi H. Distribution of bacteriocin genes in the lineages of *L. plantarum.* Sci Rep 2021; 11(1).

[45]   Franz CMAP, Du Toit M, Olasupo NA, Schillinger U, Holzapfel WH. Plantaricin D, a bacteriocin produced by *Lactobacillus plantarum* BFE 905 from ready-to-eat salad. Lett Appl Microbiol 1998; 26(3): 231-5.
[http://dx.doi.org/10.1046/j.1472-765X.1998.00332.x] [PMID: 9569716]

[46]   Hu M, Zhao H, Zhang C, Yu J, Lu Z. Purification and characterization of plantaricin 163, a novel bacteriocin produced by *Lactobacillus plantarum* 163 isolated from traditional Chinese fermented vegetables. J Agric Food Chem 2013; 61(47): 11676-82.
[http://dx.doi.org/10.1021/jf403370y] [PMID: 24228753]

[47]   Barbosa MS, Todorov SD, Ivanova IV, *et al.* Characterization of a two-peptide plantaricin produced by *Lactobacillus plantarum* MBSa4 isolated from Brazilian salami. Food Control 2016; 60: 103-12.
[http://dx.doi.org/10.1016/j.foodcont.2015.07.029]

[48]   An Y, Wang Y, Liang X, *et al.* Purification and partial characterization of M1-UVs300, a novel bacteriocin produced by *Lactobacillus plantarum* isolated from fermented sausage. Food Control 2017; 81: 211-7.
[http://dx.doi.org/10.1016/j.foodcont.2017.05.030]

[49]   Shi Y, Zhai Q, Li D, *et al.* Restoration of cefixime-induced gut microbiota changes by *Lactobacillus* cocktails and fructooligosaccharides in a mouse model. Microbiol Res 2017; 200: 14-24.
[http://dx.doi.org/10.1016/j.micres.2017.04.001] [PMID: 28527760]

[50]   Cebeci A, Gürakan C. Properties of potential probiotic *Lactobacillus plantarum* strains. Food Microbiol 2003; 20(5): 511-8.
[http://dx.doi.org/10.1016/S0740-0020(02)00174-0]

[51]   Sunanliganon C, Thong-Ngam D, Tumwasorn S, Klaikeaw N. *Lactobacillus plantarum* B7 inhibits *Helicobacter pylori* growth and attenuates gastric inflammation. World J Gastroenterol 2012; 18(20): 2472-80.
[http://dx.doi.org/10.3748/wjg.v18.i20.2472] [PMID: 22654444]

[52]   Karasu N, Şimşek Ö, Çon AH. Technological and probiotic characteristics of *Lactobacillus plantarum* strains isolated from traditionally produced fermented vegetables. Ann Microbiol 2010; 60(2): 227-34.
[http://dx.doi.org/10.1007/s13213-010-0031-6]

[53]   Arasu MV, Al-Dhabi NA, Ilavenil S, Choi KC, Srigopalram S. *In vitro* importance of probiotic *Lactobacillus plantarum* related to medical field. Saudi J Biol Sci 2016; 23(1): S6-S10.
[http://dx.doi.org/10.1016/j.sjbs.2015.09.022] [PMID: 26858567]

[54]   Rossi F, Capodaglio A, Dellaglio F. Genetic modification of *Lactobacillus plantarum* by heterologous gene integration in a not functional region of the chromosome. Appl Microbiol Biotechnol 2008; 80(1): 79-86.

[http://dx.doi.org/10.1007/s00253-008-1527-x] [PMID: 18512055]

[55]   Xu YG, Yu H, Zhang L, *et al.* Probiotic properties of genetically engineered *Lactobacillus plantarum* producing porcine lactoferrin used as feed additive for piglets. Process Biochem 2016; 51(6): 719-24.
[http://dx.doi.org/10.1016/j.procbio.2016.03.007]

[56]   Han W, Mercenier A, Ait-Belgnaoui A, *et al.* Improvement of an experimental colitis in rats by lactic acid bacteria producing superoxide dismutase. Inflamm Bowel Dis 2006; 12(11): 1044-52.
[http://dx.doi.org/10.1097/01.mib.0000235101.09231.9e] [PMID: 17075345]

[57]   Zhang Y, Kumar A, Hardwidge PR, Tanaka T, Kondo A, Vadlani PV. D-lactic acid production from renewable lignocellulosic biomass *via* genetically modified *Lactobacillus plantarum*. Biotechnol Prog 2016; 32(2): 271-8.
[http://dx.doi.org/10.1002/btpr.2212] [PMID: 26700935]

[58]   Helanto M, Kiviharju K, Leisola M, Nyyssölä A. Metabolic engineering of *Lactobacillus plantarum* for production of L-ribulose. Appl Environ Microbiol 2007; 73(21): 7083-91.
[http://dx.doi.org/10.1128/AEM.01180-07] [PMID: 17873078]

[59]   Tsuji A, Okada S, Hols P, Satoh E. Metabolic engineering of *Lactobacillus plantarum* for succinic acid production through activation of the reductive branch of the tricarboxylic acid cycle. Enzyme Microb Technol 2013; 53(2): 97-103.
[http://dx.doi.org/10.1016/j.enzmictec.2013.04.008] [PMID: 23769309]

[60]   Okano K, Uematsu G, Hama S, *et al.* Metabolic engineering of *Lactobacillus plantarum* for direct L-lactic acid production from raw corn starch. Biotechnol J 2018; 13(5): 1700517.
[http://dx.doi.org/10.1002/biot.201700517] [PMID: 29393585]

[61]   Liu S, Nichols NN, Dien BS, Cotta MA. Metabolic engineering of a *Lactobacillus plantarum* double ldh knockout strain for enhanced ethanol production. J Ind Microbiol Biotechnol 2006; 33(1): 1-7.
[http://dx.doi.org/10.1007/s10295-005-0001-3] [PMID: 16193282]

[62]   Martins EMF, Ramos AM, Vanzela ESL, Stringheta PC, De Oliveira Pinto CL, Martins JM. Products of vegetable origin: A new alternative for the consumption of probiotic bacteria. Food Res Int 2013; 51(2): 764-70.
[http://dx.doi.org/10.1016/j.foodres.2013.01.047]

CHAPTER 8

Genome Editing in *Bacillus Licheniformis*: Current Approaches and Applications

Steji Raphel[1,2] and **Prakash M. Halami**[1,2,*]

[1] *Department of Microbiology and Fermentation Technology, CSIR- Central Food Technological Research Institute, Mysuru-570020, India*

[2] *Academy of Scientific and Innovative Research (AcSIR), Ghaziabad, Uttar Pradesh, India*

Abstract: *Bacillus licheniformis* has been regarded as an exceptional microbial cell factory for the production of biochemicals and enzymes. The complete genome sequencing and annotation of the genomes of industrially-relevant *Bacillus* species has uplifted our understanding of their properties and helped in the progress of genetic manipulations in other *Bacillus* species. The genome sequence analysis has given information on the different genes and their functional importance. Post-genomic studies require simple and highly efficient tools to enable genetic manipulation. With the developments of complete genome sequences and simple genetic manipulation tools, the metabolic pathways of *B. licheniformis* could be rewired for the efficient production of interest chemicals. However, gene editing (such as gene knockout) is laborious and time consuming using conventional methods. Recently, useful tools for the genetic engineering of *Bacillus* species have emerged from the fields of systems and synthetic biology. The recent progress in genetic engineering strategies as well as the available genetic tools that have been developed in *Bacillus licheniformis* species, has conveniently enabled multiple modifications in the genomes of *Bacillus* species and thereby improved its use in the industrial sector.

Keywords: *Bacillus licheniformis*, Complete genome sequence, CRISPR-Cas9, Gene editing, Genome engineering tools.

INTRODUCTION

The finding of genes as the information carrier made it the focus of experimental research and investigation for its applications in mankind and their betterment. The discovery of DNA along with the hereditary role of genes, leads to the research on the methods to manipulate site-specific editing in the genome [1]. The engineering of nucleases could create site-specific double-strand breaks in DNA,

---

* **Corresponding author Prakash M. Halami:** Department of Microbiology and Fermentation Technology, CSIR-Central Food Technological Research Institute, Mysuru–570020, India & Academy of Scientific and Innovative Research (AcSIR), Ghaziabad, Uttar Pradesh, India; E-mail: *prakashalami@cftri.res.in*

which stimulated the homologous recombination [2]. From then, the genome editing field showed a tremendous change from a niche research technique to a universal genetic research tool. In genome editing, natural cellular pathways were exploited to repair DNA breaks introduced in the genome by various types of nucleases like zinc finger nucleases (ZFNs), transcription activator-like effector nucleases (TALENs) and engineered meganucleases [3]. These are the old techniques used to produce specific DNA breaks in the desired DNA sequence. The CRISPR-Cas system is one of the latest nuclease platforms, which is now being used extensively for gene editing [4]. This technique is based on the adaptive immune mechanism shown by bacteria and archaea, where it uses RNA guided nucleases to degrade the specific sequence of the viral DNA [5]. The Cas9 endonuclease (Cas9-gRNA ribonucleoprotein (RNP) complex) of this system depends on a 100 nucleotide sgRNA (single guide RNA) for cleaving the chromosomal target, which is activated by the hybridization of the sgRNA on the target DNA [6]. Because of its specificity, speed and cost effectiveness, it is the most efficient and simple technique in the field of gene editing [7].

The latest applications of genome editing, along with its latest tools mentioned above, made it a keen area of interest for researchers. Such advanced gene editing tools have made it possible that any gene can be deleted or incorporated into the genome of any organism. It also allows system level identification and exposing the functionality of the genome along with direct editing and modulation of the DNA functions of any organism of choice [8]. The advancement in this area created a revolution in the food and biopharmaceutical industries and also can cure innumerable genetic diseases. Apart from its clinical use, it can also be used for the production of recombinant therapeutic protein products by engineering yeast and CHO cells [9].

Besides the enormous therapeutic applications of gene editing in eukaryotes, it can also achieve specific deletions, insertions and point mutations in the bacterial genome, making it relevant in the industrial sector. Such metabolic engineering in bacterial cells can manipulate their cellular functions to achieve an increased yield and productivity of the targeted value-added biochemicals [10]. The ease and efficacy in the genetic manipulation of *E. coli* species are attributable to its extensive application in research and industrial fields [11]. This made it one of the most extensively used cellular factories for the making of industrially important enzymes and biochemicals [12].

Recently, the species *Bacillus* also received improved importance because of their application in genome editing and for the synthesis of heterologous proteins, vitamins, antibiotics, valuable enzymes and chemicals [13]. The *Bacillus* species, which are non-pathogenic and have a history of secure exploitation in foods, are being used in fermentation and industrial level production. From the global and gene specific level regulation and modification of metabolic pathways, numerous cellular phenotypes can be obtained in *Bacillus* species. For such genomic alteration of *Bacillus*, various valuable tools have been recently developed, which helped realize the potential of this bacterium [14].

The whole genome sequence data of *Bacillus* species helped in understanding the metabolic pathways along with supplying an overview of protein machinery. The enhanced understanding of the strains allowed advances in genetic manipulations in the same and related species [13]. And thus, efficient and simple genetic tools emerged from the fields of systems and synthetic biology to enable multiple modifications in the bacterial genome conveniently. In view of this, the current chapter is discussing the progress in the development of genetic tools and the strategies for the genetic modification in *Bacillus* species with an emphasis on *Bacillus licheniformis*.

CHARACTERISTICS OF *BACILLUS* SPP.

The genus *Bacillus* comprises rod-shaped, chemo heterotrophic, Gram-positive bacteria which have peritrichous flagella for their motility and have no capsules. They are catalase positive aerobic or facultative anaerobic organisms [15]. They are distinguished by their ability to produce dormant endospores aerobically when they are grown in unfavourable conditions. These spores can be round, cylindrical, oval or kidney shaped. They are dormant, rigid, non-reproductive structures with high resistance to heat, cold, disinfectants, desiccation, radiation and drying and can remain for a longer period [16].

They are majorly free –living non-parasitic saprophytes, which are seen ubiquitously in soil but are also isolated from other environments like water, air, food and vegetables and human and animal gut [17]. Though some of the *Bacillus* species are pathogenic in nature to humans and animals, they have a crucial role in balancing the ecological systems. They are the most heterogenous group in terms of phenotypic and genotypic characteristics. They typically exhibit large, flat colonies on non-selective media and are often beta-hemolytic [18].

The genus *Bacillus* is classified taxonomically in the family Bacillaceae, order Bacillales, class Bacilli, phylum Firmicutes, and domain Bacteria. Representing 1 of the 27 genera within the family Bacillaceae, the genus *Bacillus* comprises more than 140 recognized species, including nonpathogenic saprophytes, vertebrate pathogens, and invertebrate parasites [19].

The significance of *Bacillus* species relies on their ability to synthesize antibiotics/metabolites, which shows antagonistic effects on pathogenic microbes

[15]. Moreover, these are being used in medicine and the pharmaceutical industry to control various diseases in humans, animals and plants. The ability of *Bacillus* species to produce a wide variety of metabolites with antimicrobial activity lead to the use of such species as biological control agents [20]. Since they are the most effective sources of antibiotics, their non-pathogenic, non-toxic (though not all) and inexpensive attributes made them an attractive candidate for commercial production and also for controlling harmful microorganisms [21].

The potential probiotic functionalities of several *Bacillus* strains have been screened in several *in vitro* and *in vivo* models. Like the other probiotics, *Bacillus* also exhibits characteristics such as inhibition of pathogens, antibiotics [22] and bacteriocins [23] production and inhibition of the expression of virulence genes (quorum quenching) [24]. They compete with pathogenic microbes for adhesion sites and thereby inhibit their growth. They also can lyse the cell wall of pathogenic microbes by producing lytic enzymes like chitinases, cellulases, proteases and β-1,3-glucanases [25]. They can enhance their growth by providing nutrients and also by enzymatic digestion through the production of digestive enzymes. Another characteristic of *Bacillus* species is that they can enhance the host's innate and adaptive immunity using their immunostimulatory effects and by the stimulation of beneficial gut microflora [26]. *B. subtilis, B. licheniformis, B. pumilus* and *B. amyloliquefaciens* are some of the *Bacillus* species used mainly as probiotics.

The role of *Bacillus* species in industrial fermentation processes has become very crucial. They are known for the production of a range of commercial products like chemicals, antibiotics, surfactants, food enzymes and vitamins [27 - 29].

Among the different *Bacillus* species, *Bacillus licheniformis* is an important industrial microorganism because of its high-yield fermentation capacity in the production of industrial enzymes, biopolymers and valuable metabolites like amino acids, peptides, lipopeptides, nucleic acids and phospholipids [30, 31]. *B. licheniformis,* along with its closely related *B.subtilis* species, together contributes to nearly half of the global enzyme industry output. Table **1** lists the key industrial fermentation processes that use *Bacillus licheniformis*. The products listed are produced by different strains of *B. licheniformis,* which are industrially relevant.

This spore-forming saprophytic organism is widely distributed in the environment. Though most of the other bacilli are predominantly aerobic, *Bacillus licheniformis* is a facultative anaerobe which allows it to grow in additional ecological niches. Also, specific isolates of *B. licheniformis* are known for their denitrification capacity [32].

Table 1. Industrial fermentation products from *Bacillus licheniformis*.

Product	Industrial Application	Strain used	References
Specialty Chemicals			
Poly-γ-glutamic acid	Food, Cosmetics, Medicine	*Bacillus licheniformis* CCRC12826	[35]
Enzymes			
α- Amylase	Food, Paper, Starch, Textile, Brewing	*Bacillus licheniformis* MTCC 1483	[36]
Penicillinase	Pharmaceutical industries	*Bacillus licheniformis* 749/C	[32]
Xylanase	Food	*Bacillus licheniformis* strains P11, A99	[32]
Cycloglucosyltransferase	Food, starch	*Bacillus licheniformis* DSM13	[32]
β- mannanase	Food & feed, Pulp & paper	*Bacillus licheniformis* TJ-101	[37]
Alkaline phosphatase	Detergent	*Bacillus licheniformis* MTCC 1483	[38]
Alkaline (serine-) protease	Detergent	*Bacillus licheniformis* NCIM 2042	[39]
Biofuels and Bio-Based Chemicals			
Polyhydroxybutyrate	Polymers	*Bacillus licheniformis* MSBN12	[40]
H_2	Biofuel	*Bacillus licheniformis* JK-1	[41]
Peptides			
Bacitracin	Antibiotic	*Bacillus licheniformis* DW2	[42]
Proticin	Antibiotic	*Bacillus licheniformis* FH-G-439	[32]
Plant probiotic (Endophyte)	Agriculture	*Bacillus licheniformis* strains B003, BSK-4	[43]
Animal probiotic	Animal feed	*Bacillus licheniformis* strain 1.183	[43]
Aqua probiotic	Aquaculture feed	*Bacillus licheniformis* DHAB1	[18]

The bacterium is considered under the 'generally regarded as safe' (GRAS) designation for its use in the food industry. It has abundant enzyme systems, requires simple nutrition and is known for its ability to secrete products into the extracellular medium, which makes it a potential platform for biomanufacturing [33]. The recent advancement in biotechnology and bioinformatics, eventually led to the genomic level understanding of *Bacillus licheniformis* which helped in the comprehension and analysis of the different strains [34].

On account of the genetic miscellany and robustness of the bacterium, it has been considered as a valuable host in the industrial production of many beneficial chemicals, proteins and other antibiotics. Still, the application of many of such strains in the industrial field is hindered owing to its low transformation efficiency. With the development in the whole genome sequencing and genetic engineering tools, the metabolic pathways of the bacterium could be modified in order to make the interested products in an efficient manner [44].

GENOMICS OF *BACILLUS LICHENIFORMIS*

Complete genome sequences of industrially important organisms reveal its potential use in industry, human health and in the environment. It also gives an understanding on the bacterial evolution regarding their adaptation strategies to overcome the competitors. Through whole genome sequencing, reference sequence information of multiple organisms can be achieved. Also the genomic variations associated with phenotypes can be identified.

Several species of *Bacillus* have been completely sequenced for the further exploitation of these bacteria in industrial processes. It also gave knowledge of their biochemistry, physiology and genetics. *Bacillus* is one of the best-studied genera in the genomic database as more than 108 complete and draft genomes of *Bacillus* are accessible till now.

Given that *Bacillus licheniformis* is an industrial organism, functional and comparative genomics studies of this species can give insights into the techniques where the strain can be improved for industrial purposes as well as for a better understanding of genome evolution [45].

General Genomic Features

The genome of *B. licheniformis* consists of a circular chromosome with an average size of 4.2 Mbp, where it can be upto 4.5 Mbp in the case of *B. licheniformis* S127 strain. The average G+C content for the species ranges from 45% to 47%. It usually does not consist of any plasmids. Using several gene-finding programs, more than 4200 genes have been found, while in the strain BL010, 5341 genes are present. The protein coding sequences (CDSs) constitute almost 87% of the total genome. Among 4286 genes present in the genome of the DSM13 strain, 4223 are predicted as CDSs, while 4705 predicted CDSs are present in the genome of the BL010 strain. On the chromosome, these CDSs are oriented primarily in the direction of replication. Among a total of 1318 proteins found in the *B. licheniformis* genome with hits in the PIR database, 1106 are hypothetical proteins, while the remaining 212 are conserved hypothetical proteins. Mainly, the genome consists of 72 tRNA genes representing all the 20 amino acids where some strains carry 81 genes for the same. They mostly carry 7 rRNA operons with 24 rRNA genes and some with 21 rRNA genes. An exception can be seen with *Bacillus licheniformis* S127 strain which contains 36 rRNA genes. The details of genomes of various *Bacillus licheniformis* strains and their comparative analysis are depicted in Table **2**.

Table 2. Comparative analysis of genome features of *Bacillus licheniformis* strains.

Sr. No.	Microorganism	Genome Size (Mbp)	GC(%)	Total no. of Genes	No.of CDSs	No.of tRNAs	No.of rRNAs	Refs.
1	*Bacillus licheniformis* DSM13	4.22	46.2	4286	4223	72	21	[46]
2	*Bacillus licheniformis* BL010	4.28	47.0	5341	4705	81	24	[47]
3	*Bacillus licheniformis* 9945A	4.37	45.9	4359	4226	72	21	[48]
4	*Bacillus licheniformis* 14ADL4	4.33	45.86	4419	3900	81	24	[49]
5	*Bacillus licheniformis* 0DA23-1	4.40	45.96	4518	4104	81	24	[50]
6	*Bacillus licheniformis* S127	4.5	45.5	4923	4806	81	36	[51]
7	*Bacillus licheniformis* CBA7126	4.21	46.24	-	4276	81	24	[52]

Metabolic Activity and Extracellular Enzymes

Bacillus licheniformis secretome is found to encode a number of enzymes that hydrolyze polysaccharides, lipids, proteins and nutrients. The presence of genes encoding enzymes like endoglucanase, β-mannanase, cellulose-1,4-β-cellobiosidase and β-glucosidase collectively degrades and utilizes cellulose as a carbon and energy source. Apart from this, it also contains numerous carbohydrase encoding genes such as xylanase, endo-arabinase and pectate lyase, which help the bacterium to grow under varying oxygen conditions and also for starch hydrolysis [46].

The organism produces a repertoire of extracellular proteases for the utilization of nitrogenous compounds. It includes serine proteases (*aprE, epr, vpr*), metalloprotease (*mpr*), and exo- and endopeptidases (*yjbG, ydiC, gcp, ykvY, ampS, yusX, ywaD, pepT*). The apr gene encodes Subtilisin Carlsberg, which is the majorly secreted alkaline serine protease. It also possesses genes encoding arginase, asparaginase and glutaminase enzymes for utilizing amino and imino nitrogen from respective amino acids. Genes for nitrogen assimilation and transport are also present in the genome. The key genes involved are *glnA, glnR, tnrA* and *nrgA,* along with *nasABC,* which is for the nitrate transport [32].

The higher yield of the industrially important compound poly-γ-glutamic acid (γ-PGA) by *B. licheniformis* WX-02 strain has been explained by Yangste *et al*. by the comparative genomic analysis of the strain with the type strain ATCC 14580. And they reported that the difference in the comP gene in the gene cluster comQXPA required for the regulation of γ-PGA is the reason for the higher production of the compound by the WX-02 strain [53].

Export Proteins

The bacterium makes use of the Sec and Tat pathways for the export of the enzymes to the extracellular space. The genes involve *secA, secD, secE, secF, secY, ffh* and *ftsY* that encodes the proteins of the sec pathway [54] while *tatAY, tatCD* and *tatCY* are involved in the tat pathway export. Also, genes encoding lipoprotein signal peptide cleaving peptidases and sortases are present for the export of proteins to the cell surface [32].

Secondary Metabolites

Bacillus species are well known for their ability to synthesise a range of secondary metabolites. A wide variety of pathways are involved in their synthesis, and the environmental conditions, as well as the genetic makeup of the producing microbe, affect their activity. The bioactive natural products produced by *Bacillus* are primarily low molecular weight polypeptides which show antibiotic activity [55].

Many *Bacillus licheniformis* strains produce Bacitracin, a non-ribosomally synthesized cyclic peptide antibiotic. Though the presence of Bacitracin synthase genes varies with strains, 50% of the strains of the species are found to carry the bac operon. *B. licheniformis* ATCC14580 strain is devoid of this operon, but it carries an operon for lichenysin biosynthesis, which is a biosurfactant with antimicrobial activity. Also, some strains are found to carry a 14- gene cluster encoding the lantibiotic, and lichenicidin. It is a broad spectrum antibacterial compound against gram positive bacteria with LanT, LanM and Lan A proteins present on the biosynthesis cluster [56].

Apart from the presence of antibiotic synthesis genes, antibiotic resistance genes are also present in certain strains of *B. licheniformis*. The strain 14ADL4 is reported to have genes that confer lincomysin resistance. Genes like *lmrA* and *lmrB* are the two significant genes which are found to provide resistance to lincomycin as well as to lincosamide antibiotics like clindamycin. Also, in some strains, the resistance to clindamycin and erythromycin is attributable to the presence of the erythromymcin ribosome methylase gene (*erm*) [49]. The chromosomes of many strains are found to carry putative resistance genes *aph* and *aadK* for streptomycin resistance and a chloramphenicol acetyltransferase protein encoding cat gene conferring chloramphenicol resistance [57].

Stress Response

The stress response of the strain varies with the type of stress and thus is determined by the upregulation or downregulation of specific genes associated with it. In glucose starved cells, induction of genes like *acoA, B* and *L, acuA, bglH* and *rocD* can be seen, while the induction of *glnA, trpA* and *D* is specific for nitrogen straved cells. In phosphate starved cells, *phoP, pstBA* and *S, ylaK* and *ydhD* genes are reported to get upregulated [58].

In oxidative stress, upregulation of *sufB* and *C* genes are reported, and in the case of osmotic stress, an elevation in the synthesis of stress proteins *ClpC, E* and *P* along with *gbsAB* gene expression can be seen. High salinity is also found to induce the gene clusters encoding a transport system belonging to the ABC super family of exporters [59]. Among the different stress proteins, drought stress induced a higher expression of the specific genes, including *Cadhn, VA, sHSP* and *CaPR-10,* which was reported based on the study on *B. licheniformis* K11 strain [60].

GENOME EDITING IN *BACILLUS LICHENIFORMIS*

Bacillus licheniformis is a promising industrial host strain due to its GRAS status as well as its ability to secrete a higher number of enzymes and proteins to the extracellular medium [61]. But it has many undesirable properties due to its habitant on soil and plant rhizosphere. The discontinuous supply of nutrients, abiotic stress and competition from other microbes has posed challenges in their environment and thus adopted sporulation as well as the secretion of viscous substances in larger amounts for survival in such environments and also to increase the cellular competitiveness [58, 62]. Similar adverse conditions, like oxidative and osmotic stress, starvation *etc.,* are confronted by bacteria in industrial fermentation processes and thus lead to incomplete sterilization, decreased enzyme yield and higher production costs in industrial operations [63]. To overcome this, the most commonly used effective method is the deletion of such undesirable intrinsic genes, which can improve the traits in the engineered strains. Such modifications in *B. licheniformis* cells without any undesirable properties are necessary for its use in industrial applications [64].

Apart from this, certain gene regulation and expression methods have been used as host-modification strategies. However, the extremely low transformation efficiency of *B. licheniformis* along with the lack of efficient genetic manipulation techniques and vector availability, created challenges in the molecular level research on the organism, which hindered the development of engineered strains and their use in the industrial sector [65]. So, strategies were developed to improve the strain using better promoters, strong signal peptides and by optimizing the fermentation conditions through genome editing and metabolomic studies. For this reason, researchers had to rely on engineered endonucleases and other genome editing tools for the optimization of the strains, which satisfied the requirements making them acceptable in industrial usage. But some of these methods are less efficient and costly and involve complex manipulations. Thus, it became a necessity of the current scenario to develop economical and practical genome editing tools which are highly efficient and easy to operate for obtaining improved strains of *B. licheniformis* [34].

GENOME EDITING TOOLS DEVELOPED IN *BACILLUS LICHENIFORMIS*

A wide range of numerous systems have been developed for the genetic manipulation and genomic engineering of the industrial workhorse *B. licheniformis* for its betterment and, thereby widespread usage in different sectors.

Inducible Promoter Based Genome Engineering

Various inducible promoter systems have been developed for the analysis of gene expression and function in *B. licheniformis* strains. It make use of inducing agents like hormones, heat shock, regulatory factors *etc.,* for manipulating gene expression.

The proteases and amylases produced by *B. licheniformis* are growth associated enzymes which depend on the substrates in the medium. This mechanism is undesirable in industrial processes, as the strains should secrete a maximum amount of enzymes using minimum resources [66]. Inducible expression systems are a solution for this problem as it separates enzyme production from growth. This system is being used predominantly in *E.coli* and *Saccharomyces cerevisiae* for enzyme production. Only a few functional inducible promoters are reported in *Bacillus* species, and among them, the xylose promoter is used because of its high level transcription initiation capacity [67].

Few studies are there in which the xylose operon is used for the expression of individual genes in *B. licheniformis*. For the inducible expression of the trehalose synthase gene, three inducible expression systems were constructed in *B.*

licheniformis using xylose operons from *B. subtilis, B. licheniformis* and *B. megaterium*. The system was constructed in a way that allows only strictly controlled and high-level expression of the trehalose synthase gene [68]. This system has also been used for the secretory expression of maltogenic alpha amylase (BLMA) under strict conditions in *B. licheniformis*. Mutations in the highly conserved nucleotides in the catabolite responsive element (cre) of the promoter reduced the carbon catabolite repression (CCR) and thereby induced the BLMA expression [69]. The xylose inducible promoter, along with a constitutive promoter and surfactin operon promoter, has been reported for its use in the improvement of lychenisin production in *B. licheniformis* strains [70].

Apart from this, an ammonia-inducible promoter has been used in the organism for its conventional use in fermentations. This glucose independent promoter is found to mediate enzyme expression and thus reported to enhance the enzyme yield [71].

An auto-inducible expression system, *phyL* promoter was also demonstrated in *B. licheniformis*. It uses phytate as the inducer of this system and also is an alternative phosphate source for this bacterium. This expression system is suitable for the overexpression of target genes in industrial batch fermentation processes of *B. licheniformis* under growth conditions [72].

Another inducible expression system developed in *B. licheniformis* is the mannitol operon based inducible system. The two promoters, *PmtlA* and *PmtlR* involved in the sugar alcohol uptake pathway were characterized and developed as regulated promoters for this system. Along with sugar alcohol, mannitol, mannose, sorbitol, sorbase and arabinose also are reported as inducers for this expression system [73].

Pyrimidine Metabolism Based Genome Engineering

Genome modification approaches involve a vital genetic tool, counterselection markers. It uses an *in vitro* modified allele for the targeted and unmarked replacement of a gene. It exploits the genes for the metabolism of purine or pyrimidine and is based on the fact that their analogs can be converted to toxic compounds. The selection of clones lacking the gene encoding the purine or pyrimidine converting enzyme is by plating them on media containing their analogs. So, the strains devoid of purine or pyrimidine biosynthesis genes are used as the parental strain for genome modification. Efficient and commonly used counterselection marker being used in *Bacillus* species includes *upp* encoding Uracil Phospho Ribosyl Transferase (UPRTase) and *pyrF* encoding Orotidine 5' –

Mono Phosphate decarboxylase (OMPdecase) [14]. The Fig. (**1**) depicts the pyrimidine metabolism pathway and the mechanism of action of pyrimidine analogs.

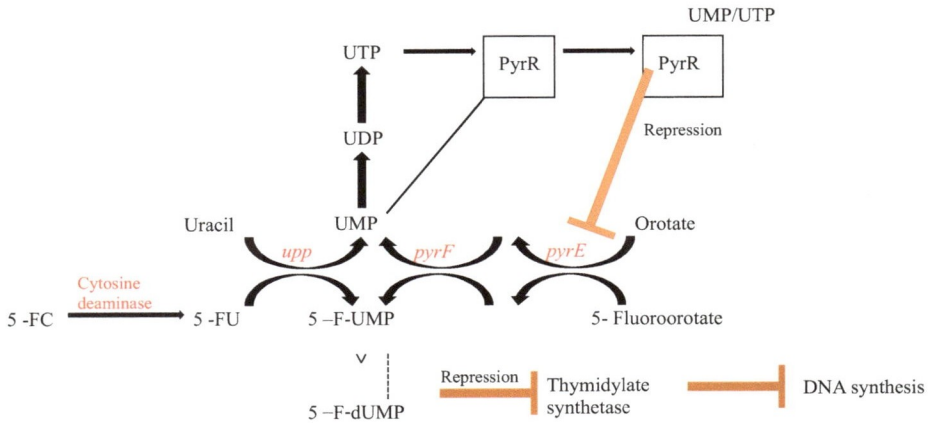

Fig. (1). Pyrimidine metabolism pathway and mechanism of action of pyrimidine analogs . *pyrE* and *pyrF* encode OPRTase (Orotate phosphoribosyl transferase) and OMPdecase (Orotidine monophosphate decarboxylase), respectively, which convert orotate to UMP (Uridine monophosphate) as well as 5-fluoro orotate to 5-fluoro UMP. 5-fluoru cytosine gets converted to 5-fluoro uracil by cytosine deaminase. The UPRTase (Uracil phosphoribosyl transferase) encoded by *upp* converts this 5-FU and Uracil to 5-F-UMP and UMP respectively. 5-F-UMP will get further metabolized into 5-F-dUMP which is a toxic metabolite that represses the thymidylate synthetase formation and thereby inhibits DNA synthesis.

A convenient counterselection strategy was developed in *B. licheniformis* for the generation of markerless chromosomal deletions. The system was reported to use *codBA* as a counterselection marker and 5-fluorocytosine as the counter selective compound. It was used to understand the role of two genes encoding the proteins, similar to alanine dehydrogenase, which are involved in the amino acid metabolism [74].

Transconjugation Based Plasmid System

A gene modification method without natural competence and marker was developed in *B. licheniformis* species by Rachinger *et al* [75]. The pKVM series of conjugative shuttle vectors containing the flanking regions of the target gene were used for the chromosomal gene deletion. These shuttle vectors carry the origin of replication from temperature-sensitive pE194ts and a thermostable β-galactosidase. This allowed the selection of recombinant clones on X-ga-

-containing agar plates using the blue/white screening method. Also, it can be conjugated to *B. licheniformis* and *B. subtilis* strains. Homologous recombination allowed the target locus integration of the vector, and its subsequent excision. Furthermore, appropriate selection markers were used to identify the recombinant cells. Such pKVM vectors have applications in creating efficient deletions and insertions in *B. licheniformis* and other *Bacillus* strains.

Site Specific Recombination Techniques

A recombinase that catalyzes the recombination between two site specific recognition sites is the basis for the site specific recombination (SSR) systems. A targeted DNA integration, deletion or inversion can be achieved using this system. Different SSR systems are being used in microbial gene editing which includes Cre/loxP system, Xis/attP system, FLP/FRT system, *etc*. However, the most accepted and widely used one is the CRISPR/Cas system because of its high efficiency and ease of manipulation. Multiple sequential mutations on the same chromosome can be attained with SSR systems owing to their higher recombination efficiency [14].

FLP/FRT System

Zongwen *et al*. constructed an FLP/FRT gene editing system in *B. licheniformis* and verified it by knocking out of two genes along with a single gene knock in. It can repeatedly use a single selectable marker for multiple mutations. The knockout plasmids carrying respective homology arms, resistance genes and, FRT sites were transformed into the bacteria, where the target genes were replaced *via* twice homologous recombination with the respective deletion cassette. Later, the resistance marker was excised by introducing an FLP recombinase reading frame through an expression plasmid. A knock-in plasmid was also constructed in order to expand the practicability of this system [76].

CRISPR/Cas Systems

In the genome editing field, the CRISPR-Cas9 system recently received widespread acceptance as it is easy to operate, highly efficient and requires short manufacturing time [77]. This Clustered Regularly Interspaced Short Palindromic Repeats (CRISPR) system is a kind of natural acquired immune system present in many bacteria and archaea to prevent the invasion of foreign DNA [78]. Type II-A CRISPR Cas9 system of *Streptococcus pyogens* is the commonly used tool in genome editing. It typically comprises major two components, a Cas9 DNA endonuclease and a chimeric single-guided RNA (sgRNA), which reprogram the target site to recruit the Cas9 protein [79]. Thereby, the endonuclease creates a double-strand break at the target site. The cleavage site will then get repaired by a non-homologous end joining (NHEJ) or homology-directed repair (HDR) mechanism.

The CRISPR-Cas9 mediated genome editing in bacteria has two categories:

1) Cas9 mediated gene editing: This system is found to have application in genome editing, disease treatments and also in epigenome studies. Double gene disruption, inactivation of bacterial genes and deletion of major DNA fragments can be achieved with the plasmid-based CRISPR-Cas9 system and NHEJ. Nevertheless, it poses certain limitations as it is inappropriate for certain bacterial species because of its less expression or lack of key NHEJ components like Ku protein, ATP-dependent DNA ligase, and DNA polymerase LigD. This can lead to cleavage and, thus bacterial death [80, 81].

2) Cas9 nickase (Cas9n)-mediated genome editing: This system is useful to overcome the above-mentioned limitations. Cas9n combined with HDR is believed as a more defined means of genome editing as it modifies only the fragment between the donor templates. Therefore, it can be used to avoid the lethality in bacteria which are deficient in NHEJ components. It precisely edits the genome outside the target sequence and thus decreases the incidence of off-target mutations [82].

Cas9 Mediated Genome Editing

For the industrially important, poorly transformable strain *B. licheniformis* 2709, an inducible single-plasmid CRISPR/Cas9 genome editing tool was developed. The shuttle vector pWH1520 was selected to construct the knockout plasmids for this genome editing. It consisted of a pS promoter and a strong pLY-2 promoter for the expression of Cas9 endonuclease and the transcription of a single guide RNA, respectively. This could improve the genome editing efficiency of the strain and thereby facilitate the metabolic modification of this potent industrial strain. This system was verified by engineering the strain by mutating two significant genes (*chiA* and *amyl*), which encode the secreted enzymes. The study reported production of a 24.8% higher yield of alkaline protease by the mutant (BL Δchi Δamy) than the parent strain. This suggested the key role of the developed genome editing tool in the production and synthesis of important industrial enzymes and other valuable bioactive compounds [34].

In another study, the researchers suggested the development of a non-mucoid strain of *B. licheniformis* using the CRISPR-Cas9 system. The secretion of mucoid exopolymers by the bacteria can hinder the strain's development as well as its metabolite production. To overcome this, the *pgsBCAE* operon encoding polyglutamate synthase and the *sacB* gene encoding levansucrase were deleted using CRISPR-Cas9 and thereby generated a non-mucoid strain. Also, the genes *ldhA, dgp* and *adhE* encoding lactate dehydrogenase, D-α-glycerophosphatase, and alcohol dehydrogenase, respectively, were deleted and generated a byproduct-free strain. The study demonstrated that the development of these strains can be used for the mass production of 2,3-butanediol (BDO) using a naturally isolated *B. licheniformis* strain [83].

Similarly, CRISPR-Cas9 mediated homologous recombination system was used for engineering *B. licheniformis* 4071-15 strain for the stereospecific production of (2R, 3S) - BDO isomer. An engineered triple mutant strain was obtained by the sequential deletion of the target genes, *ldh, dgp* and *acoR,* by repeated one-step cycles. This resulted in high efficiency in strain exceeding 60%. Also, the *gdh* gene encoding the (2R,3R)-butanediol dehydrogenase was deleted for the production of (2R, 3S) – BDO. The resulting mutant strain was reported to produce 115 g/L of (2R, 3S) – BDO in 64 hours by a two-stage fed batch fermentation process [84].

A conditional CRISPR/Cas9 system was constructed in *B. licheniformis* with the Cas9 gene under the control of a xylose-inducible promoter. Also, this 4107 bp-Cas9 gene was incorporated into a thermosensitive shuttle vector. This system could significantly improve the transformation ratio of the strain from less than 0.1 to 2.42 cfu/μg DNA owing to the fact that the expression of the Cas9 gene could be repressed without xylose. This conditional design has two advantages over the commonly used constitutive expression pattern in the CRISPR/Cas9 system. First, a silent Cas9 gene without induction improves the success rate when *B. licheniformis* is transformed with the all-in-one plasmid of a large size. Second, efficient genome editing can be triggered by the inducer xylose, which increases flexibility in the control of DNA nuclease dosage. Thereby, this system could significantly improve the success rate of genome editing in the host and could contribute to the efficient usage of *B. licheniformis* for various industrial purposes [11].

Cas9 Nickase Mediated Genome Editing

The study carried out by Li *et al.* on *B. licheniformis* DW2 developed an efficient, quick and organized genome editing technique based on the CRISPR-Cas9 nickase system with overexpression driven by the P43 promoter. The CRISPR-Cas9n technique was used for deleting the *yvmC* gene and attained 100% efficiency in the strain. The study found that, this system could effectively execute single-gene deletion, multiple-gene disruption, large DNA fragment deletion, and single-gene integration. It was verified by deleting six protease genes along with the bacitracin synthase gene cluster *bacABC* and produced the

DWc9nΔ7 strain which achieved a significant increase of 25.7% nattokinase activity. This paved insights into the genome editing methods that can be applied in *B. licheniformis* for the development of the strain in the future [85].

CRISPR Interference System

The CRISPR interference (CRISPRi) is derived from CRISPR/Cas9 which is an efficient means for gene suppression. The CRISPRi system consists of two components: a nuclease-deficient Cas9 (dCas9) with DNA binding capability and a single-guide RNA (sgRNA). The dCas9/sgRNA complex can prevent transcription of the targeted gene by the definite sgRNA. This multiplexed system can be used to inhibit multiple genes simultaneously. The CRISPRi system constructed in *B. licheniformis* was validated using several endogenous genes, including *yvmC, cypX, alsD, pta, ldh*, and essential gene *rpsC,* where simultaneous repression of multiple genes with high efficiency was observed. CRISPRi was engaged to change carbon flux into L-valine synthesis in this study, where a targeted repression of the genes *alsD* and *bcd* encoding acetolactate decarboxylase and leucine dehydrogenase, respectively, was achieved. This combinatorial repression blocked the by-product formation and prevented L-valine degradation, resulting in a significant increase in L-valine titer. This suggested that the CRISPRi system can be served as an excellent genetic reprogramming tool for functional genomics and metabolic engineering studies of *B. licheniformis* [44].

APPLICATIONS OF GENOME EDITING TECHNIQUES

With the emergence of highly resourceful genome editing tools, it became very helpful for researchers to introduce site-specific modifications in the genomes of a variety of cell types and organisms rapidly and economically. The core technologies such as CRISPR-Cas9, ZFNs, TALENs *etc.,* which are developed to ease genome editing, have revolutionized this field and accelerated scientific innovations and discoveries in the areas of human gene therapy, synthetic biology, drug discovery, disease modelling, agricultural sciences and neurosciences [86].

Genetically modified and metabolically engineered bacterial and yeast strains can be generated for synthetic biology by means of targeted nucleases kind of gene editing tools. Certain bacterial species are considered the most vital sources of industrially relevant secondary metabolites. In recent years, CRISPR-Cas9 has been proven to be an important technique having the ability to generate engineered bacterial strains with improved metabolite production capacity.

Bacillus licheniformis, as discussed earlier, is considered as an inevitable industrial host because of its exceptional protein synthesis and secretion capacity.

This bacteria has successfully been used for the overexpression and commercial production of a tremendous range of industrially relevant products. Numerous proficient methods have been developed for the overexpression of a specific gene in such strains by gene editing, deletion or disruption. Also, established site-specific recombination methods based on homologous recombination and selective chromosomal amplification are employed for target gene amplification. Certain methods like Cre/lox, Xer/dif, FLP/FRT and CRISPR/Cas9 systems have been modified and effectively used for target gene expression in such selected organisms [87].

The study carried out by Wang, Shiyi, *et al.* [88] provided an ideal approach as well as a promising *B. licheniformis* strain for high level production of the cyclodipeptide, pulcherriminic acid. CRISPR/Cas9 nickase mediated genome editing in *B. licheniformis* DW2 strain produced a new strain, DWc9n*. This strain was employed for the efficient production of pulcherriminic acid using a multistep metabolic engineering approach. It was achieved by cyclodipeptide overexpression and transport enhancement and also by redirecting the carbon flux towards Leucine and leucyl-tRNA biosynthesis, which are the precursors.

The same genetically modified strain, *B. licheniformis* DWc9n* was also engineered for the high-level production of short branched- chain fatty acids (SBCFAs), which are multifunctional platform chemicals used in various fields. A productive endogenous pathway for SBCFAs *via* CRISPR/Cas9n was achieved by overexpressing the SBCFAs synthesis genes by replacing the native promoter of the bkd operon. This resulted in an increased titre of SBCFAs that could be efficiently produced from waste materials like bean dreg and crude glycerol [89].

The work carried out by Shen *et al* [87] developed a novel strategy for the production of a genetically stable thermophilic a-amylase overexpressing *B. licheniformis* strain using a single generation cycle. To force the complete loss of the replicative plasmid while the integration and amplification of non-replicable expression plasmid in the chromosome, an RNase encoded by mazF from *Escherichia coli* was used as a counter-selection marker. This method can also be used for developing strains that can overexpress industrially important enzymes with higher efficiency.

The probiotic properties of *B. licheniformis* can also be improved using such genome engineering approaches. The gene profiling and metabolic engineering strategies targeting specific genes of the particular strains can be beneficial in modifying the organism for its enhanced probiotic traits as well as the production yields. With these methods, we can also identify a potential probiotic strain possessing an effective mechanism of action in the gut. Single nucleotide changes made by CRISPR Cas9 based gene editing can provide a chance to improve the scope of probiotics in maintaining health issues and its applications in other related fields [90].

The diverse bacterial genome engineering techniques recognize the biological significance of the genomic data sets. It may offer a novel thought to develop the industrial and commercial production of delectable biological agents. Regardless of the successes already attained, countless disputes remain before the full potential of genome editing can be recognized. The very first is the development of novel tools which are able to introduce genomic modifications in the absence of DNA breaks. As the frequency of off-target modifications can be directly proportional to the duration of cellular exposure to a nuclease, self-inactivating vectors can improve the specificity of genome editing [86]. Only when we consider and accept such challenges posed by these techniques, then will genome editing technologies truly be able to be exploited to their full extend.

CONCLUSION

The industrial workhorse *Bacillus licheniformis* has become an essential platform for the production of various enzymes and chemicals. Regulation and modification of several metabolic pathways at global and gene-specific level can attain a range of cellular phenotypes in the *Bacillus* species. The importance of *Bacillus* species as production hosts for manufacturing commodities was established by the advancements in the genetic engineering strategies.This made it a competitive strain along with *E. coli* and *S. cerevisiae,* which are considered as the traditional industrial microbes.

In recent years, many valuable tools for genetic modification of *Bacillus* species have been developed. Here, we have summarized the current genetic engineering strategies and their recent progress in the genome editing field is discussed. Also, the different genome editing tools developed in *Bacillus licheniformis* for the betterment of the strain are evaluated. However, these strategies also posed particular challenges and limitations, thus, a broad range of applications and proficient tools for systems-level genomic editing are still in need.

ACKNOWLEDGEMENTS

The authors would like to thank the Director, CSIR- Central Food Technological Research Institute, for the facilities provided. SR would like to thank the Council for Scientific and Industrial Research for granting Senior Research Fellowship. We acknowledge Dr. Sreedhar RV for his critical comments on the manuscript.

REFERENCES

[1]   Doudna JA, Charpentier E. The new frontier of genome engineering with CRISPR-Cas9. Science 2014; 346(6213): 1258096.
[http://dx.doi.org/10.1126/science.1258096] [PMID: 25430774]

[2]   Porteus MH, Baltimore D. Chimeric nucleases stimulate gene targeting in human cells. Science 2003; 300(5620): 763.
[http://dx.doi.org/10.1126/science.1078395] [PMID: 12730593]

[3]   Baker M. Gene-editing nucleases. Nat Methods 2012; 9(1): 23-6.
[http://dx.doi.org/10.1038/nmeth.1807] [PMID: 22312637]

[4]   Cong L, Ran FA, Cox D, *et al.* Multiplex genome engineering using CRISPR/Cas systems. Science 2013; 339(6121): 819-23.
[http://dx.doi.org/10.1126/science.1231143] [PMID: 23287718]

[5]   Gasiunas G, Barrangou R, Horvath P, Siksnys V. Cas9–crRNA ribonucleoprotein complex mediates specific DNA cleavage for adaptive immunity in bacteria. Proc Natl Acad Sci 2012; 109(39): E2579-86.
[http://dx.doi.org/10.1073/pnas.1208507109] [PMID: 22949671]

[6]   Auer TO, Duroure K, De Cian A, Concordet JP, Del Bene F. Highly efficient CRISPR/Cas9-mediated knock-in in zebrafish by homology-independent DNA repair. Genome Res 2014; 24(1): 142-53.
[http://dx.doi.org/10.1101/gr.161638.113] [PMID: 24179142]

[7]   Kanchiswamy CN, Maffei M, Malnoy M, Velasco R, Kim JS. Fine-tuning next-generation genome editing tools. Trends Biotechnol 2016; 34(7): 562-74.
[http://dx.doi.org/10.1016/j.tibtech.2016.03.007] [PMID: 27167723]

[8]   Hsu PD, Lander ES, Zhang F. Development and applications of CRISPR-Cas9 for genome engineering. Cell 2014; 157(6): 1262-78.
[http://dx.doi.org/10.1016/j.cell.2014.05.010] [PMID: 24906146]

[9]   Gupta SK, Shukla P. Gene editing for cell engineering: Trends and applications. Crit Rev Biotechnol 2017; 37(5): 672-84.
[http://dx.doi.org/10.1080/07388551.2016.1214557] [PMID: 27535623]

[10]   Cho S, Shin J, Cho BK. Applications of CRISPR/Cas system to bacterial metabolic engineering. Int J Mol Sci 2018; 19(4): 1089.
[http://dx.doi.org/10.3390/ijms19041089] [PMID: 29621180]

[11]   Li Y, Wang H, Zhang L, *et al.* Efficient genome editing in *Bacillus licheniformis* mediated by a conditional CRISPR/Cas9 system. Microorganisms 2020; 8(5): 754.
[http://dx.doi.org/10.3390/microorganisms8050754] [PMID: 32429599]

[12]   Ajikumar PK, Xiao WH, Tyo KEJ, *et al.* Isoprenoid pathway optimization for Taxol precursor overproduction in *Escherichia coli.* Science 2010; 330(6000): 70-4.
[http://dx.doi.org/10.1126/science.1191652] [PMID: 20929806]

[13]   Liu L, Liu Y, Shin H, *et al.* Developing *Bacillus* spp. as a cell factory for production of microbial enzymes and industrially important biochemicals in the context of systems and synthetic biology. Appl Microbiol Biotechnol 2013; 97(14): 6113-27.
[http://dx.doi.org/10.1007/s00253-013-4960-4] [PMID: 23749118]

[14]   Dong H, Zhang D. Current development in genetic engineering strategies of *Bacillus* species. Microb Cell Fact 2014; 13(1): 63.
[http://dx.doi.org/10.1186/1475-2859-13-63] [PMID: 24885003]

[15]   Amin M, Rakhisi Z. Isolation and identification of bacillus species from soil and evaluation of their antibacterial properties. Avicenna J Clin Microbiol Infect 2015; 2(1): 23233.

[16]   Gordon RE, Haynes WC, Pang CH. The genus Bacillus Agricultural research service. US Department

of Agriculture 1973.

[17]   Tidjiani Alou M, Rathored J, Khelaifia S, *et al*. *Bacillus rubiinfantis* sp. nov. strain mt2T, a new bacterial species isolated from human gut. New Microbes New Infect 2015; 8: 51-60.
         [http://dx.doi.org/10.1016/j.nmni.2015.09.008] [PMID: 27076912]

[18]   Kuebutornye FKA, Abarike ED, Lu Y. A review on the application of *Bacillus* as probiotics in aquaculture. Fish Shellfish Immunol 2019; 87: 820-8.
         [http://dx.doi.org/10.1016/j.fsi.2019.02.010] [PMID: 30779995]

[19]   Elshaghabee FMF, Rokana N, Gulhane RD, Sharma C, Panwar H. *Bacillus* as potential probiotics: Status, concerns, and future perspectives. Front Microbiol 2017; 8: 1490.
         [http://dx.doi.org/10.3389/fmicb.2017.01490] [PMID: 28848511]

[20]   Silo-Suh LA, Lethbridge BJ, Raffel SJ, He H, Clardy J, Handelsman J. Biological activities of two fungistatic antibiotics produced by *Bacillus cereus* UW85. Appl Environ Microbiol 1994; 60(6): 2023-30.
         [http://dx.doi.org/10.1128/aem.60.6.2023-2030.1994] [PMID: 8031096]

[21]   Abarike ED, Cai J, Lu Y, *et al*. Effects of a commercial probiotic BS containing *Bacillus subtilis* and *Bacillus licheniformis* on growth, immune response and disease resistance in Nile tilapia, *Oreochromis niloticus*. Fish Shellfish Immunol 2018; 82: 229-38.
         [http://dx.doi.org/10.1016/j.fsi.2018.08.037] [PMID: 30125705]

[22]   Pinchuk IV, Bressollier P, Verneuil B, *et al. In vitro* anti-*Helicobacter pylori* activity of the probiotic strain *Bacillus subtilis* 3 is due to secretion of antibiotics. Antimicrob Agents Chemother 2001; 45(11): 3156-61.
         [http://dx.doi.org/10.1128/AAC.45.11.3156-3161.2001] [PMID: 11600371]

[23]   Chu F, Kearns DB, Branda SS, Kolter R, Losick R. Targets of the master regulator of biofilm formation in *Bacillus subtilis*. Mol Microbiol 2006; 59(4): 1216-28.
         [http://dx.doi.org/10.1111/j.1365-2958.2005.05019.x] [PMID: 16430695]

[24]   Musthafa KS, Saroja V, Pandian SK, Ravi AV. Antipathogenic potential of marine *Bacillus* sp. SS4 on N-acyl-homoserine-lactone-mediated virulence factors production in *Pseudomonas aeruginosa* (PAO1). J Biosci 2011; 36(1): 55-67.
         [http://dx.doi.org/10.1007/s12038-011-9011-7] [PMID: 21451248]

[25]   Biziulevièius GA, Þukaitë V. Comparative antimicrobial activity of lysosubtilin and its acid-resistant derivative, Fermosorb. Int J Antimicrob Agents 2002; 20(1): 65-8.
         [http://dx.doi.org/10.1016/S0924-8579(02)00117-6] [PMID: 12127714]

[26]   Makrinos DL, Bowden TJ. Natural environmental impacts on teleost immune function. Fish Shellfish Immunol 2016; 53: 50-7.
         [http://dx.doi.org/10.1016/j.fsi.2016.03.008] [PMID: 26973022]

[27]   Van Dijl JM, Hecker M. *Bacillus subtilis*: From soil bacterium to super-secreting cell factory. Microb Cell Fact 2013; 12(1): 3.
         [http://dx.doi.org/10.1186/1475-2859-12-3] [PMID: 23311580]

[28]   Eggersdorfer M, Laudert D, Létinois U, *et al*. One hundred years of vitamins : A success story of the natural sciences. Angew Chem Int Ed 2012; 51(52): 12960-90.
         [http://dx.doi.org/10.1002/anie.201205886] [PMID: 23208776]

[29]   Singh P, Patil Y, Rale V. Biosurfactant production: Emerging trends and promising strategies. J Appl Microbiol 2019; 126(1): 2-13.
         [http://dx.doi.org/10.1111/jam.14057] [PMID: 30066414]

[30]   Guo J, Zhang H, Wang C, Chang JW, Chen LL. Construction and analysis of a genome-scale metabolic network for *Bacillus licheniformis* WX-02. Res Microbiol 2016; 167(4): 282-9.
         [http://dx.doi.org/10.1016/j.resmic.2015.12.005] [PMID: 26776566]

[31]   Saadi S, Saari N, Anwar F, Abdul Hamid A, Ghazali HM. Recent advances in food biopeptides:

Production, biological functionalities and therapeutic applications. Biotechnol Adv 2015; 33(1): 80-116.
[http://dx.doi.org/10.1016/j.biotechadv.2014.12.003] [PMID: 25499177]

[32]   Rey MW, Ramaiya P, Nelson BA, *et al.* Complete genome sequence of the industrial bacterium *Bacillus licheniformis* and comparisons with closely related *Bacillus* species. Genome Biol 2004; 5(10): r77.
[http://dx.doi.org/10.1186/gb-2004-5-10-r77] [PMID: 15461803]

[33]   Bashir F, Asgher M, Hussain F, Randhawa MA. Development and characterization of cross-linked enzyme aggregates of thermotolerant alkaline protease from *Bacillus licheniformis*. Int J Biol Macromol 2018; 113: 944-51.
[http://dx.doi.org/10.1016/j.ijbiomac.2018.03.009] [PMID: 29510168]

[34]   Zhou C, Liu H, Yuan F, *et al.* Development and application of a CRISPR/Cas9 system for *Bacillus licheniformis* genome editing. Int J Biol Macromol 2019; 122: 329-37.
[http://dx.doi.org/10.1016/j.ijbiomac.2018.10.170] [PMID: 30401651]

[35]   Shih IL, Van YT, Chang YN. Application of statistical experimental methods to optimize production of poly(γ-glutamic acid) by *Bacillus licheniformis* CCRC 12826. Enzyme Microb Technol 2002; 31(3): 213-20.
[http://dx.doi.org/10.1016/S0141-0229(02)00103-5]

[36]   Kaur PS, Kaur S, Kaur H, Sondhi S. Statistical optimization of the production of alpha-amylase from *Bacillus licheniformis* MTCC 1483 using paddy straw as substrate. J Commer Biotechnol 2017; 23(2).
[http://dx.doi.org/10.5912/jcb784]

[37]   Liu ZH, Qi W, He ZM. Optimization of β-mannanase production from *Bacillus licheniformis* TJ-101 using response surface methodology. Chem Biochem Eng Q 2008; 22(3): 355-62.

[38]   Pandey SK, Banik RM. Extractive fermentation for enhanced production of alkaline phosphatase from *Bacillus licheniformis* MTCC 1483 using aqueous two-phase systems. Bioresour Technol 2011; 102(5): 4226-31.
[http://dx.doi.org/10.1016/j.biortech.2010.12.066] [PMID: 21227688]

[39]   Bhunia B, Dey A. Statistical approach for optimization of physiochemical requirements on alkaline protease production from *Bacillus licheniformis* NCIM 2042. Enzyme research 2012; 2012.

[40]   Sathiyanarayanan G, Saibaba G, Seghal Kiran G, Selvin J. Process optimization and production of polyhydroxybutyrate using palm jaggery as economical carbon source by marine sponge-associated *Bacillus licheniformis* MSBN12. Bioprocess Biosyst Eng 2013; 36(12): 1817-27.
[http://dx.doi.org/10.1007/s00449-013-0956-9] [PMID: 23670633]

[41]   Kalia VC, Jain SR, Kumar A, Joshi AP. Fermentation of biowaste to H2 by *Bacillus licheniformis*. World J Microbiol Biotechnol 1994; 10(2): 224-7.
[http://dx.doi.org/10.1007/BF00360893] [PMID: 24420953]

[42]   Wang Q, Zheng H, Wan X, *et al.* Optimization of inexpensive agricultural by-products as raw materials for bacitracin production in *Bacillus licheniformis* DW2. Appl Biochem Biotechnol 2017; 183(4): 1146-57.
[http://dx.doi.org/10.1007/s12010-017-2489-1] [PMID: 28593603]

[43]   Mongkolthanaruk W. Classification of *Bacillus* beneficial substances related to plants, humans and animals. J Microbiol Biotechnol 2012; 22(12): 1597-604.
[http://dx.doi.org/10.4014/jmb.1204.04013] [PMID: 23221520]

[44]   Zhan Y, Xu Y, Zheng P, *et al.* Establishment and application of multiplexed CRISPR interference system in *Bacillus licheniformis*. Appl Microbiol Biotechnol 2020; 104(1): 391-403.
[http://dx.doi.org/10.1007/s00253-019-10230-5] [PMID: 31745574]

[45]   Sharma A, Satyanarayana T. Comparative genomics of Bacillus species and its relevance in industrial microbiology. Genomics insights 2013; 6: GEI-S12732.

[46]   Veith B, Herzberg C, Steckel S, *et al.* The complete genome sequence of *Bacillus licheniformis* DSM13, an organism with great industrial potential. J Mol Microbiol Biotechnol 2004; 7(4): 204-11.
[PMID: 15383718]

[47]   He S, Feng K, Ding T, *et al.* Complete genome sequence of *Bacillus licheniformis* BL-010. Microb Pathog 2018; 118: 199-201.
[http://dx.doi.org/10.1016/j.micpath.2018.03.037] [PMID: 29578060]

[48]   Rachinger M, Volland S, Meinhardt F, Daniel R, Liesegang H. First insights into the completely annotated genome sequence of *Bacillus licheniformis* strain 9945A. Genome Announc 2013; 1(4): e00525-13.
[http://dx.doi.org/10.1128/genomeA.00525-13] [PMID: 23908277]

[49]   Jeong DW, Lee B, Lee JH. Complete genome sequence of *Bacillus licheniformis* 14ADL4 exhibiting resistance to clindamycin. Microbiolog Soc Korea 2018; 54(2): 169-70.

[50]   Jeong DW, Lee B, Heo S, Jang M, Lee JH. Complete genome sequence of *Bacillus licheniformis* strain 0DA23-1, a potential starter culture candidate for soybean fermentation. Kor J Microbiol 2018; 54(4): 453-5.

[51]   Ostrov I, Sela N, Freed M, *et al.* Draft genome sequence of *Bacillus licheniformis* S127, isolated from a sheep udder clinical infection. Genome Announc 2015; 3(5): e00971-15.
[http://dx.doi.org/10.1128/genomeA.00971-15] [PMID: 26430024]

[52]   Lee C, Kim JY, Song HS, *et al.* Genomic analysis of *Bacillus licheniformis* CBA7126 isolated from a human fecal sample. Front Pharmacol 2017; 8: 724.
[http://dx.doi.org/10.3389/fphar.2017.00724] [PMID: 29081747]

[53]   Yangtse W, Zhou Y, Lei Y, *et al.* Genome sequence of *Bacillus licheniformis* WX-02. J Bacteriol 2012; 194(13): 3561-2.
[http://dx.doi.org/10.1128/JB.00572-12] [PMID: 22689245]

[54]   Kunst F, Ogasawara N, Moszer I, *et al.* The complete genome sequence of the Gram-positive bacterium *Bacillus subtilis.* Nature 1997; 390(6657): 249-56.
[http://dx.doi.org/10.1038/36786] [PMID: 9384377]

[55]   Stoica RM, Moscovici M, Tomulescu C, *et al.* Antimicrobial compounds of the genus *Bacillus*: A review. Rom Biotechnol Lett 2019; 24(6): 1111-9.
[http://dx.doi.org/10.25083/rbl/24.6/1111.1119]

[56]   Dunlap CA, Kwon SW, Rooney AP, Kim SJ. *Bacillus paralicheniformis* sp. nov., isolated from fermented soybean paste. IntJ systemevolutmicrobiol 2015; 65(Pt_10): 3487-92.

[57]   Agersø Y, Bjerre K, Brockmann E, *et al.* Putative antibiotic resistance genes present in extant *Bacillus licheniformis* and *Bacillus paralicheniformis* strains are probably intrinsic and part of the ancient resistome. PLoS One 2019; 14(1): e0210363.
[http://dx.doi.org/10.1371/journal.pone.0210363] [PMID: 30645638]

[58]   Voigt B, Schroeter R, Schweder T, *et al.* A proteomic view of cell physiology of the industrial workhorse *Bacillus licheniformis.* J Biotechnol 2014; 191: 139-49.
[http://dx.doi.org/10.1016/j.jbiotec.2014.06.004] [PMID: 25011098]

[59]   Schroeter R, Hoffmann T, Voigt B, *et al.* Stress responses of the industrial workhorse *Bacillus licheniformis* to osmotic challenges. PLoS One 2013; 8(11): e80956.
[http://dx.doi.org/10.1371/journal.pone.0080956] [PMID: 24348917]

[60]   Lim JH, Kim SD. Induction of drought stress resistance by multi-functional PGPR *Bacillus licheniformis* K11 in pepper. Plant Pathol J 2013; 29(2): 201-8.
[http://dx.doi.org/10.5423/PPJ.SI.02.2013.0021] [PMID: 25288947]

[61]   Zhou C, Zhou H, Zhang H, Lu F. Optimization of alkaline protease production by rational deletion of sporulation related genes in *Bacillus licheniformis.* Microb Cell Fact 2019; 18(1): 127.

[http://dx.doi.org/10.1186/s12934-019-1174-1] [PMID: 31345221]

[62]   Yi G, Liu Q, Lin J, Wang W, Huang H, Li S. Repeated batch fermentation for surfactin production with immobilized *Bacillus subtilis* BS-37: two-stage pH control and foam fractionation. J Chem Technol Biotechnol 2017; 92(3): 530-5.
[http://dx.doi.org/10.1002/jctb.5028]

[63]   Bressuire-Isoard C, Broussolle V, Carlin F. Sporulation environment influences spore properties in *Bacillus*: Evidence and insights on underlying molecular and physiological mechanisms. FEMS Microbiol Rev 2018; 42(5): 614-26.
[http://dx.doi.org/10.1093/femsre/fuy021] [PMID: 29788151]

[64]   Zhou C, Zhou H, Li D, Zhang H, Wang H, Lu F. Optimized expression and enhanced production of alkaline protease by genetically modified *Bacillus licheniformis* 2709. Microb Cell Fact 2020; 19(1): 45.
[http://dx.doi.org/10.1186/s12934-020-01307-2] [PMID: 32093734]

[65]   Liu Y, Xu Y, Fan S, Bo J, Wang J, Lu F. Study on the influencing conditions in the electro-transformation efficiency. In Proceedings of the 2012 International Conference on Applied Biotechnology,. (ICAB 2012) (p. 1845).

[66]   Banerjee A, Leang C, Ueki T, Nevin KP, Lovley DR. Lactose-inducible system for metabolic engineering of *Clostridium ljungdahlii*. Appl Environ Microbiol 2014; 80(8): 2410-6.
[http://dx.doi.org/10.1128/AEM.03666-13] [PMID: 24509933]

[67]   Bhavsar AP, Zhao X, Brown ED. Development and characterization of a xylose-dependent system for expression of cloned genes in *Bacillus subtilis*: Conditional complementation of a teichoic acid mutant. Appl Environ Microbiol 2001; 67(1): 403-10.
[http://dx.doi.org/10.1128/AEM.67.1.403-410.2001] [PMID: 11133472]

[68]   Li Y, Gu Z, Zhang L, Ding Z, Shi G. Inducible expression of trehalose synthase in *Bacillus licheniformis*. Protein Expr Purif 2017; 130: 115-22.
[http://dx.doi.org/10.1016/j.pep.2016.10.005] [PMID: 27751933]

[69]   Li Y, Jin K, Zhang L, Ding Z, Gu Z, Shi G. Development of an inducible secretory expression system in *Bacillus licheniformis* based on an engineered xylose operon. J Agric Food Chem 2018; 66(36): 9456-64.
[http://dx.doi.org/10.1021/acs.jafc.8b02857] [PMID: 30129762]

[70]   Qiu Y, Xiao F, Wei X, Wen Z, Chen S. Improvement of lichenysin production in *Bacillus licheniformis* by replacement of native promoter of lichenysin biosynthesis operon and medium optimization. Appl Microbiol Biotechnol 2014; 98(21): 8895-903.
[http://dx.doi.org/10.1007/s00253-014-5978-y] [PMID: 25085615]

[71]   Shen P, Niu D, Permaul K, Tian K, Singh S, Wang Z. Exploitation of ammonia-inducible promoters for enzyme overexpression in *Bacillus licheniformis*. J Ind Microbiol Biotechnol 2021; 48(5-6): kuab037.
[http://dx.doi.org/10.1093/jimb/kuab037] [PMID: 34124759]

[72]   Trung NT, Hung NM, Thuan NH, Canh NX, Schweder T, Jürgen B. An auto-inducible phosphate-controlled expression system of *Bacillus licheniformis*. BMC Biotechnol 2019; 19(1): 3.
[http://dx.doi.org/10.1186/s12896-018-0490-6] [PMID: 30626366]

[73]   Xiao F, Li Y, Zhang Y, *et al.* Construction of a novel sugar alcohol-inducible expression system in *Bacillus licheniformis*. Appl Microbiol Biotechnol 2020; 104(12): 5409-25.
[http://dx.doi.org/10.1007/s00253-020-10618-8] [PMID: 32333054]

[74]   Kostner D, Rachinger M, Liebl W, Ehrenreich A. Markerless deletion of putative alanine dehydrogenase genes in *Bacillus licheniformis* using a codBA-based counterselection technique. Microbiology (Reading) 2017; 163(11): 1532-9.
[http://dx.doi.org/10.1099/mic.0.000544] [PMID: 28984230]

[75] Rachinger M, Bauch M, Strittmatter A, *et al.* Size unlimited markerless deletions by a transconjugative plasmid-system in *Bacillus licheniformis.* J Biotechnol 2013; 167(4): 365-9.
[http://dx.doi.org/10.1016/j.jbiotec.2013.07.026] [PMID: 23916947]

[76] Li Z, Li Y, Gu Z, *et al.* Development and verification of an FLP/FRT system for gene editing in *Bacillus licheniformis.* Chin J Biotechnol 2019; 35(3): 458-71.
[PMID: 30912354]

[77] Zhang H, Cheng QX, Liu AM, Zhao GP, Wang J. A novel and efficient method for bacteria genome editing employing both CRISPR/Cas9 and an antibiotic resistance cassette. Front Microbiol 2017; 8: 812.
[http://dx.doi.org/10.3389/fmicb.2017.00812] [PMID: 28529507]

[78] Barrangou R, Marraffini LA. CRISPR-Cas systems: Prokaryotes upgrade to adaptive immunity. Mol Cell 2014; 54(2): 234-44.
[http://dx.doi.org/10.1016/j.molcel.2014.03.011] [PMID: 24766887]

[79] Jinek M, Chylinski K, Fonfara I, Hauer M, Doudna JA, Charpentier E. A programmable dual-RNA–guided DNA endonuclease in adaptive bacterial immunity. science 2012; 337(6096): 816-21.

[80] Xu T, Li Y, Van Nostrand JD, He Z, Zhou J. Cas9-based tools for targeted genome editing and transcriptional control. Appl Environ Microbiol 2014; 80(5): 1544-52.
[http://dx.doi.org/10.1128/AEM.03786-13] [PMID: 24389925]

[81] Su T, Liu F, Gu P, *et al. el al.* A CRISPR-Cas9 assisted non-homologous end-joining strategy for one-step engineering of bacterial genome. Sci Rep 2016; 6(1): 37895.
[http://dx.doi.org/10.1038/srep37895] [PMID: 28442746]

[82] Standage-Beier K, Zhang Q, Wang X. Targeted large-scale deletion of bacterial genomes using CRISPR-nickases. ACS Synth Biol 2015; 4(11): 1217-25.
[http://dx.doi.org/10.1021/acssynbio.5b00132] [PMID: 26451892]

[83] Song CW, Rathnasingh C, Song H. CRISPR-Cas9 mediated metabolic engineering of a mucoid *Bacillus licheniformis* isolate for mass production of 2,3-butanediol. Biochem Eng J 2021; 175: 108141.
[http://dx.doi.org/10.1016/j.bej.2021.108141]

[84] Song CW, Rathnasingh C, Park JM, Kwon M, Song H. CRISPR-CAS9 mediated engineering of *Bacillus licheniformis* for industrial production of (2R, 3S)-butanediol. Biotechnol Prog 2021; 37(1): e3072.
[http://dx.doi.org/10.1002/btpr.3072] [PMID: 32964665]

[85] Li K, Cai D, Wang Z, He Z, Chen S. Development of an efficient genome editing tool in *Bacillus licheniformis* using CRISPR-Cas9 nickase. Appl Environ Microbiol 2018; 84(6): e02608-17.
[http://dx.doi.org/10.1128/AEM.02608-17] [PMID: 29330178]

[86] Gaj T, Sirk SJ, Shui S, Liu J. Genome-editing technologies: Principles and applications. Cold Spring Harb Perspect Biol 2016; 8(12): a023754.
[http://dx.doi.org/10.1101/cshperspect.a023754] [PMID: 27908936]

[87] Shen P, Niu D, Liu X, *et al.* Overexpression of thermophilic α-amylase in *Bacillus licheniformis* using a high efficiency chromosomal integration and amplification strategy. Res Sq 2021.

[88] Wang S, Wang H, Zhang D, *et al.* Multistep metabolic engineering of *Bacillus licheniformis* to improve pulcherriminic acid production. Appl Environ Microbiol 2020; 86(9): e03041-19.
[http://dx.doi.org/10.1128/AEM.03041-19] [PMID: 32111589]

[89] Shi J, Zhan Y, Zhou M, *et al.* High-level production of short branched-chain fatty acids from waste materials by genetically modified *Bacillus licheniformis.* Bioresour Technol 2019; 271: 325-31.
[http://dx.doi.org/10.1016/j.biortech.2018.08.134] [PMID: 30292131]

[90]   Yadav R, Kumar V, Baweja M, Shukla P. Gene editing and genetic engineering approaches for advanced probiotics: A review. Crit Rev Food Sci Nutr 2018; 58(10): 1735-46.
[http://dx.doi.org/10.1080/10408398.2016.1274877] [PMID: 28071925]

SUBJECT INDEX

www.ingramcontent.com/pod-product-compliance
Lightning Source LLC
Chambersburg PA
CBHW041701210326
41598CB00007B/489